THE STING OF THE WILD

THE STING OF THE WILD

JUSTIN O. SCHMIDT

Johns Hopkins University Press • Baltimore

Printed in the United States of America on acid-free paper

Johns Hopkins Paperback edition, 2018
2 4 6 8 9 7 5 3 1

Johns Hopkins University Press
2715 North Charles Street
Baltimore, Maryland 21218-4363
www.press.jhu.edu

The Library of Congress has cataloged the hardcover edition of this book as follows:
Schmidt, Justin O., 1947– author.
The sting of the wild / Justin O. Schmidt.
pages cm
Includes bibliographical references and index.
ISBN 978-1-4214-1928-2 (hardcover : alk. paper)
ISBN 978-1-4214-1929-9 (electronic)
ISBN 1-4214-1928-9 (hardcover : alk. paper)
ISBN 1-4214-1929-7 (electronic)
1. Poisonous arthropoda. 2. Insect pests. I. Title.
QL434.4′5.S36 2016
595.7′165—dc23 2015026989

A catalog record for this book is available from the British Library.

ISBN-13: 978-1-4214-2564-1
ISBN-10: 1-4214-2564-5

Special discounts are available for bulk purchases of this book. For more information, please contact Special Sales at 410-516-6936 or specialsales@press.jhu.edu.

Johns Hopkins University Press uses environmentally friendly book materials, including recycled text paper that is composed of at least 30 percent post-consumer waste, whenever possible.

To Debbie and Li

CONTENTS

PREFACE

Let's saddle up and embark on an adventure. This adventure will be in our minds, not in the thorns, mosquitoes, and sweat of the real world. Cheating? Perhaps. But don't all adventures take place in our minds as well as in our bodies? The adventure starts a million and half years ago in the savannas and open woodlands of Africa. Our band of 25 is relaxing at the base of a kopje, a large rock hill protruding from the open land. A few experienced men sit atop the kopje, scanning the landscape for potential threats and opportunities. Maybe there are some nearby lions to avoid, or a leopard lurking in a tree next to our path to the water hole, or maybe half a dozen raiders of a neighboring band are heading our way. Today is a good day with none of those threats; instead, we spy a just-born baby giraffe that appears unable to walk. Alert all boys over the age of 5 to accompany and learn from the experienced men how to seize this opportunity to feed the group a valuable treat of fresh meat. Three of the strongest men sprint-lope toward the distant giraffe to secure it from hyenas or lions that might also take note. Fortunately, no lions or hyenas had arrived on the scene, only circling jackals awaiting scraps after a kill.

Two older men and an older boy rally the eager younger boys to follow and assist the experienced men ahead of them. As we hurry along through the grass, we remember that lions, leopards, and hyenas are not the only threats. Hey, what is that sinuous movement just to the left? Snake! Caution. Years ago an adventurous youngster was bitten by a snake, and he died. Snakes are dangerous. Maybe some snakes are not dangerous. But enough snakes are dangerous that all must be feared and avoided—better safe than sorry. That problem avoided, the trip continues. We pass under a baobab tree. Ouch, what stung me? Run, a beehive is located in the tree, and the bees are unhappy. Bees, too, can be a threat. We note the location and alert the advance men, who have driven off the mother giraffe with rocks and threatening sticks and are now butchering the young giraffe to bring back to the group. After returning to camp, the boys lead the excited men with their smoking torches to the beehive location and then stand back to watch. Two main honey hunters climb the baobab with their torches to chase away and rob the bees of their precious honey and combs of brood. The rest of us have learned that bees, like snakes, are inherently frightening and dangerous, and caution is required if we are to survive and be healthy.

Our adventure is imaginary, but its message is real. Animals that can kill or hurt us are to be feared and treated with caution or avoided outright. For millions of years, our ancestors, whether human or protohuman, encountered dangerous animals, some large and powerful, some smaller but deadly, and some tiny but also able to inflict harm. These dangerous encounters indelibly etched innate fear of potentially harmful animals in our genes. We carry those genes with us to this day.

In this book, I hope to share a love of the natural world and the beauty of all forms of life. Each animal has its own inherently interesting story to tell. Each story is awaiting our attention to be told. I was fortunate to be able to explore many adventures with some of the most beautiful and fascinating insects on earth—stinging insects. Stinging insects provide a wondrous insight into a variety of lifestyles and solutions to their own day-to-day survival challenges.

The book is organized in two general sections. The first section comprising chapters 1 to 5 provides the background and theory that enhance the following chapters. The second, and larger, part is a series of chapters that go in-depth into one particular small group of insects. The reader is encouraged to skip around and read chapters in any order, as each chapter is intended to stand mainly on its own. Start at the beginning, or pick a favorite and jump in.

I have a confession to make. The writing is littered with numerical superscripts. These are references to the sources of information and can be and—I might argue—should be ignored unless something piques particular interest. I beg that these numbers not be too much of a nuisance. Let's go. I hope you enjoy the ride as much as I did.

This writing would not have been possible without the help of numerous people. Throughout my career, I have enjoyed many engaging and fruitful discussions about biology and stinging insects with numerous colleagues and friends, especially Steve Buchmann, Bob Jacobson, Bill Overal, Roy Snelling, Hayward Spangler, Chris Starr, and Murray Blum. Their ideas and discussions have enriched this book. I am indebted to John Alcock, Craig Brabant, Mathias Buck, Jim Cane, Joe Coelho, Eric Eaton, Kevin O'Neill, and Rolf Ziegler for providing information and fact checking for sections of the writing. In addition to providing ideas and information, Denis Brothers, Deby Cassill, Bill McGrew, Jon Harrison, Chuck Holliday, Jenny Jandt, Bob Jacobson, and Richard Wrangham critically read and helped improve various chapters. Louise Shaler, Elizabeth Taylor, and Tom Wiewandt encouraged me and aided in numerous ways to improve clarity, understanding, and presentation of the writing. Particular thanks to Margarethe Brummermann, Jillian Cowles, and Graham Wise for sharing and permissions to use their photographs.

I would be remiss not to mention my editors, especially Vincent J. Burke and the tremendously talented team at Johns Hopkins University Press. I am immensely grateful to Vincent for his unwavering encouragement, support, and toleration of me throughout the writing. Without him, neither this book nor my joy (most of the time) in writing it would have been possible. Finally, I am indebted to my personal editor, adviser, and colleague Bob Jacobson who greatly helped in all aspects of the writing from sentence structure and spelling to presentation of the manuscript.

THE STING OF THE WILD

1
STUNG

Children are born naturalists.

Children are born naturalists whose play is exploration of the environment surrounding them. Throughout most of human history, that surrounding environment was nature itself, a nature filled with sights, sounds, and smells of plants, animals, and the landscape. To the child, the ant walking in the play area or nipping at a food scrap was an object of interest. Equally interesting was the nearby flower with its intriguing bee, busy collecting nectar and pollen, and the lurking crab spider on the flower's edge. To the growing brain, these experiences were exciting and valuable. At this youngest stage of life, fear is muted. Fear is mostly learned by play experiences and from nearby parents and adults. For their part, the adults in the community realize that play and learning are crucial for the developing young mind and encourage or allow mostly unrestricted play for the first five years. Play transforms the young human into an aware, observant, analytical, and adaptable individual prepared to face the world as an experienced, functional adult. But ever watchful adults are vigilant in making the environment secure for children to explore and learn in safety.

Should a snake appear on the scene, quick action is taken to protect the children and to reinforce a preexisting fear of snakes. Over thousands of generations, as studies by Lynn Isbell and others have shown, humans developed a strong genetic fear and aversion to snakes and an instinct to avoid them. This instinct is biologically rooted.[1] Those individuals who lacked a fear of snakes, or who failed at avoiding snakes, frequently were bitten, with dire consequences, and sometimes died. Genes governing detection and fear of snakes were positively adaptive for individuals possessing the genes. Those with genes that did not confer strong detection and avoidance abilities were slowly eliminated from the gene pool.

As natural scientists, children learn to appreciate and value many elements in nature and to avoid others. By observing, formulating hypotheses, testing these hypotheses, noting the results of the tests, and repeating the process, they are engaging in science. This process comes naturally to children. No teachers are needed to instruct them in the method. Sadly, teachers later are needed to reinstill the scientific method after children grow older and have had this natural talent driven out of them. A paradox? Yes and no. Modern parents inherently sense that children love and need nature. That is why baby clothes are frequently adorned with fuzzy motifs of bumble bees or honey bees, and children's beds abound with stuffed animals, such as bears, tigers, and even sharks. Parents know these animals can all be dangerous in real life, so why encourage them as intimate parts of their children's lives? Could it be that parents know these mascots of nature encourage children's excitement, learning, and comfort?

My own early childhood in Appalachian Pennsylvania was not much different from that of many children around the world. My parents, unbeknownst to me, allowed and encouraged my exploration under their watchful eye. Frogs would be placed in pockets, mud pies made, and lightning bugs put in jars. I suspect these activities were not enjoyed by my mother, though they were tolerated, perhaps with the hope that I would outgrow them. At about five years of age, my well-being was sometimes entrusted to the care of my seven-year-old brother, my ten-year-old sister, and groups of older kids. I, as the youngest, needed to prove my worth to the group. One pleasant spring day the gang happened to notice a large mound of thatching ants from the genus *Formica*. These ants have no sting but produce copious quantities of formic acid, the most corrosive and acidic of the aliphatic organic acids, which they spray from the tips of their abdomens. They also bite. The combination of a solid bite breaking the skin and formic acid sprayed into the wound yields a sting-like pain. Some of the older boys dared me to sit on the ant mound, a challenge and an opportunity to prove myself that could not be missed. The ants swarmed over and under my britches and started biting my posterior. Up from the mound, down with the pants, and frantic brushing to get rid of them. No long-term damage was done, but I had learned an important lesson: insects can fight back. I continued with the group on to other adventures, a bit wiser and more experienced. Thus, my beginnings as an entomologist.

As children grow, their play turns to honing skills that may be needed in later life. For our ancestors, some of these skills were hunting and solving mysteries of nature. To master hunting skills, physical strength and coordination need sharpening, and nature must be observed, explored, and tested, and its mysteries probed. Today, in economically developed societies, hunting skills are less important, but the urge is still strong, especially in boys. The time-honored rural Pennsylvania tradition of declaring the first day of deer hunting season a school holiday exemplifies the modern continuation of old instincts. Old fields, fencerows, small woodlands, and streams abounded in the area of my childhood—perfect places for refining skills. Other than the dusty ball field, not much entertainment was available. Our small neighborhood group of six to eight boys, ranging four years in age span, was always on the alert for new adventures, whether we were climbing a challenging tree or discovering a bumble bee or hornet nest. I, perpetually the youngest, became a skilled tree climber, and, as the lightest, soon became the best tree climber in the group. In terms of running speed and throwing ability, I was at the bottom. One June day, as we were walking along a fencerow, an older boy discovered a baldfaced hornet nest deep inside the branches of a long-neglected apple tree struggling to produce green apples. What an opportunity.

What a challenge. If we threw rocks at the nest, would they attack? If they attacked, would we escape? If they stung, would it hurt? Mysteries. To solve these mysteries and to test predictions that we could escape unscathed, the oldest boy grabbed a rock and, with the rest of us watching warily behind him, hurled it toward the nest. Poor shot. Nothing happened. We all ran a short distance. A pecking order of bravery then emerged with each boy sequentially grabbing a rock, approaching closer, throwing it toward the hidden nest, and all of us running. The rocks all missed, resulting in only a few hornets flying out to investigate, and nobody got stung. Finally, it was my turn. I found the perfect rock, approached closer than anybody had dared, and gave my mightiest throw toward the nest. A direct hit. Half the nest fell to the ground. The gang, clustered about 15 feet behind me, had a head start, and I learned how the term "mad as hornets" originated. This time the hornets meant business, and I was the closest and the slowest runner. About all I remember beyond this point was that one hornet managed to sting the back of my neck several times. The exact number of stings eludes memory but was at least three or four. It felt like someone had repeatedly struck the back of my neck with a hot branding iron. This was my first experience with what would several decades later become a 2 on the insect-sting pain scale.

About this time, I changed my approach to stinging insects from the recipient of the experiments to the designer. I was a small, skinny kid with tiny fingers and sharp eyes for close-up objects, traits that later became perfect adaptations for my training as an entomologist. I wasn't proficient at baseball or football, and marbles, our favorite game, had been recently banned from school. I had little else to do during recess other than observe plants and tiny animals on the playground. One day I saw a honey bee on a dandelion. I had been told that they could sting, so I decided to see for myself. This time, rather than be the recipient of the test results, I decided to test the hypothesis on my teacher, who was watching over the playground. I picked up the bee and put it on my teacher's forearm. I learned that honey bees can sting, and my teacher learned that honey bees can be picked up by hand. My innocent test was without malice, though it did become the topic of discussion between my parents and the teacher whenever they met, even decades later. Lessons learned from stings are remembered a long time.

Prominent in most insect guidebooks, the "cow killer"—sometimes called the mule killer—is a frequent summer visitor to yards and parks throughout the southern United States and much of the Midwest. Nearly an inch long, covered with a soft, inviting red and black fur, the cow killer superficially resembles an oversized ant. The common name "velvet ants" for cow killers and other members of this worldwide and immensely successful family of more than 8,000 species is derived from its ant-like appearance. In reality, velvet ants are wingless female wasps. Male velvet ants are winged and look much like other wasps, albeit fuzzier and furrier. Female velvet ants easily vie for entry into the Guinness World Records for the greatest number of defenses known for any insect. First is their stinger, the longest stinger relative to body length of any of the true stinging insects. This is the group called the Aculeata and includes the stinging wasps, ants, and bees but not the parasitic wasps. The parasitic wasps differ from true stinging insects in that their stingers serve primarily to lay eggs and only secondarily to inject venom. Enhancing the effectiveness of the cow killer's stinger, which can be half the total insect length, is the insect's ability to aim it widely, so that it can sting a person or predator grasping any part of its body, whether it's the head, thorax, or abdomen. The pain is instantaneous and searing, much like sticking a red-hot glowing needle into your thumb. The thumb recoils, but not the pain, which continues unabated for 5–10 minutes before gradually easing. This is in addition to a rashy-nettly pain reminiscent of a nasty brush with stinging nettle plants alongside a path near a stream. A natural urge to rub the rashy sting area increases the pain and the itch, a combination just shy of torture.

During my graduate studies at the University of Georgia in Athens, I was called to a golf course whose operators were in a panic over a large aggregation of cicada killer wasps that had taken a fancy to some of the sand bunkers. Male cicada killers were busily flying around looking for females and challenging anything moving in their territory, including golfers. Meanwhile, numbers of beautiful, colorful cow killers, *Dasymutilla occidentalis,* were entering cicada killer burrows and looking for the wasps' young as food for the cow killers' young. I captured several of the cow killers and took them to the lab where I was analyzing their defenses. A young undergraduate student who was helping care for them decided one Friday evening to give them some honey and water. I received an urgent call about 11:30 p.m. from the campus infirmary concerning what to do for my panicked student who was stung while handling a cow killer and was frightened that he would not survive the night. About all I could do was advise that the bark was far worse than the bite and that while the sting was among the most painful known, the venom was among the least toxic known. He had no chance of dying, and after a little antihistamine and some tender loving care, the student was back in the lab the next day.

I am aware of few reports of young children being stung by a cow killer. Why would a child seeing a beautiful red velvety cow killer running across the backyard not simply pick it up? Perhaps some do and the screaming child cannot describe the source of the sting to the parent. But surely the parent would look for the source and should easily spot the culprit. The more likely reason cow killer stings to young children are so infrequently reported is that they rarely occur. Just as we instinctually notice and avoid snakes and spiders,[2,3] we instinctually notice bees, wasps, and other potentially dangerous stinging insects, including cow killers. The bright red and black coloration both attracts notice and signals pause and caution: look before you leap; watch before you touch. Contrasting patterns of red and black are classic aposematic warnings that signal would-be predators to "back off, leave me alone . . . if you do not, you will regret the consequences." Aposematic, derived from the Greek words *apo* = away, and *sematic* = signal, perfectly describes the cow killer. Its nasty sting backs up the warning, and in the case of the cow killer, additional warnings also come in the form of sound, a squeak of broad frequency range that resembles a miniature rattle of a Lilliputian rattlesnake, and an odorous chemical warning signal. The cow killer releases these warning chemicals from glands at the base of the mandibles (the insect's jaws), a blend of volatile ketone molecules that smells like fingernail polish remover. For nocturnal predators, or those with poor vision, one or both of the sound or smell warnings are memorably received.

Cow killer defenses do not stop here. In case of an actual attack, two powerful backup defenses come into play. The first is the immensely hard integument, or shell, of the cow killer, rather like a biological tank with a hard, impenetrable body armor. Cow killers are so hard that stainless steel insect pins sometimes bend without penetrating the body. Equally impressive and more biologically relevant, adult tarantula spiders are unable to penetrate cow killers with their impressive fangs, and, on feeling the vibratory squeaking, a feeling akin to a mini jackhammer against one's teeth, quickly release the cow killer. Immense leg strength is a final defense. The cow killer's box-shaped thorax, the middle of the three insect body parts, houses not powerful muscles for flight, as in most insects, but instead enormous muscles that power the legs. These powerful legs combined with the rounded, slippery body enable the insect to wrest itself free from a predatory grip and then rapidly run away and escape. Does a child or adult consciously know of these defenses? Not likely. But the signals are clear: be cautious and avoid me or you'll be sorry. The sting message is conveyed; truth is communicated.

2

THE STINGER

Petruchio: Come, come, you wasp, i'faith you are too angry.
Katherine: If I be waspish, best beware my sting.
—William Shakespeare, *The Taming of the Shrew,* ca. 1590

If a stinging insect could speak, the first words it might shout are, "Who is at my door?" This simple thought can determine life and death. Life requires growth, reproduction, and survival. Without any one of these, a species would not survive to the next generation and would not have arisen in the first place. Survival is the lack of death. Animal survival is simple in theory: fill one's stomach with nutritious food, and don't end up in anyone else's stomach. Both of these are challenges because most animal tummies are filled with plants, and plants are the world's best chemists at synthesizing a dazzling array of chemical compounds. These compounds are synthesized expressly to prevent being eaten or to outcompete other plants for light or nutrients. Predatory animals face other problems with their food, mainly finding, capturing, subduing, and consuming their prey. For the stinging insect, how not to be eaten assumes crucial importance, and herein lies the value of the sting.

The stinging insect is focused on not ending up in the stomach of the visitor at the entrance to the nest. Is that activity near the nest caused by an animal, or is it incidental to wind, to weather, or to nearby plants? If the latter, no predatory threat is present. If the former, the insect must determine whether the animal is a threat or is harmless.

The visitor at the entrance could simply be a bumbling cow or a wayfaring rhinoceros. The main threat, then, is being accidentally stepped on or having the insect's home, the nest, destroyed. If little risk of being eaten exists, then usually, no defensive action is needed. An interesting exception was described by Fritz Vollrath and Iain Douglas-Hamilton involving elephants and honey bees.[1] Elephants are not known to eat honey bees, but they eat trees, including large trees and branches housing honey bee colonies. Destruction of the bees' nest by an elephant knocking down or destroying the tree would be a serious problem. Bees mount stinging attacks against intruding pachyderms to eliminate the threat of nest destruction, by targeting the elephants' vulnerable eyes and nose, thereby driving the elephant herd away from the nest.

Not all intruders at a bee, wasp, or ant nest are vegetarians. Some are specifically seeking the nutritious inhabitants, their helpless young larvae and pupae, or any food stores, including honey, pollen, or dead/paralyzed prey within the nest. The stinging insect must now determine the least dangerous way to stop the intruding predator from continuing its mission. An ideal defense is to threaten the predator from a distance. Chris Starr, a former classmate, who is now in Trinidad, explored carefully the range of threats signaled to potential predators by paper wasps in the genus *Polistes*. *Polistes* are reluctant to fly off the nest and attempt to sting birds, mammals, or Chris. Instead, they engage in a series of increasing threats to avoid death by being bitten, smashed, or eaten: face the intruder with body raised high on legs; raise wings above the body and spread apart; flip wings up and down in a fast motion; buzz wings momentarily while remaining on the nest; flutter wings while still remaining on the nest; wave raised front legs toward intruder; curve the gaster (the part of the abdomen after the narrow waist); and fly off the nest but not toward the intruder. These threats often deter the intruder and at little cost to the wasp.[2] Wasps sting only after these threats fail.

Warning threats can take many forms, including sounds and smells that do not depend on predator vision. A variety of ants that include *Pogonomyrmex* harvester ants, leafcutter ants (*Atta*), Australian bull ants (*Myrmecia*), and bullet ants (*Paraponera*) stridulate to produce a broad-frequency squeak. All investigated velvet ants (Mutillidae) also readily sound a stridulatory squeak threat when they sense possible danger. Hornets in the genus *Vespa* have perfected mandible snapping or clicking as another effective acoustic threat. When I was in Japan in 1980, some generous students of social wasps and their professor assisted me in collecting an entire colony of the giant mandarin hornet, *Vespa mandarinia*. This enormous wasp with its brilliant orange blocky head vies for the title of the most intimidating insect on Earth. Their preferred food is the young of other hornets, social wasps, and honey bees, the adults of which they dispatch quickly and simply by crushing them with their enormous mandibles, having no need to waste precious venom in this operation. My Japanese colleagues and I presented a different threat to the mandarin hornets than their usual stinging prey. We were their predators, not their prey! After climbing into my armored bee suit, I grabbed a handheld insect net with a 6-inch handle and approached the hornet nest swinging. The students, perhaps wiser than I, ingeniously attached insect nets to long straight tree shoots and caught any hornets attacking me from the rear. The most memorable part of the ordeal was seeing and experiencing enormous wasps hovering eyeball-to-eyeball in front of me, loudly snapping their jaws. The best bee suit in the world could not allay the fear and awe generated by these threats. And these are not idle threats—a single sting can kill a rat. We were luckier than a rat and collected all the hornets and their nest with nobody getting stung.

Warning threats can be odorous chemicals. Ants are masters of chemical warfare with the added benefit that these odorous compounds serve as warning—"stay away or suffer the consequences of stings and bites." *Pogonomyrmex* harvester ants release volatile ketone compounds that smell like nail polish, and velvet ants use a nearly identical blend of ketones for the same purpose. Bullet ant chemical warnings include odors that resemble burnt garlic. Tarantula hawks (*Pepsis*) perhaps produce a most distinctive warning odor: a pungent and unaesthetic odor emanating from glands in the head. All of these odors can be released to warn intruders not to attack, thereby reducing the risk of actual attack. If predators attack, these odors enhance predator learning that attacking this stinging insect is punishing and a bad idea.

The insect world is a blizzard of odors. Chemical odors are not just aposematic warning signals, a minor role in insect life. Odors run most of life, ranging from sex pheromones to help males and females find one another at the right time and other pheromones communicating alarm, aggregation, and individual recognition, to a near limitless variety of chemicals that convey information about food. From early in my youth, odors that convey information about dangers, especially the presence of large, dangerous predators, have been particularly interesting to me. To sting and drive off a predator, stinging insects must first detect and recognize a predator. I worked for many years with honey bees, asking, among other questions, how honey bees detect a predator. My research revealed that odor—in this case, mammalian breath—is the strongest cue to honey bees of the presence of a mammalian predator. Breath is hot, humid, and contains carbon dioxide plus a variety of small volatile aldehydes, ketones, alcohols, esters, and other compounds. To bees, breath is a smelly pool of airborne chemicals, an immediately recognizable stench. When Africanized ("killer") honey bees arrived in Arizona in 1993, a colleague, perhaps naively, hived a few reproductive swarms of killer bees and kept them on the research location in the middle of Tucson. They would periodically go on a rampage stinging nearby people. Fortunately, none of the people were innocent public bystanders. The unfortunate victims were high school and college students, employed part time by the director of the Tucson Bee Lab, who bore the brunt of the attacks. I, as a behaviorist, needed to observe activities directly at the hive entrance and took advantage of these colonies hived by my colleague. The solution to close-range observation was simple: be "invisible" to the bees. To achieve invisibility in the presence of bees, stop breathing (granted, it is hard to stop breathing entirely for long) and move slowly. Hold your breath as you stand inches to the side of the landing board and then turn your head to exhale gently a few feet behind the hive between breaths. One day my good friend John Lewis in the maintenance department was walking 25 feet away from the colony I was watching and got stung. "Hey, Schmidt, how come I get stung and you poke your nose six inches from the entrance and don't get stung?" No, it was not bad living, just his "bad" breath.

Suppose a bee recognizes a potential predator but threats do not work? As a last resort, it may engage in the riskiest defense, planting the sting (sometimes interchangeably called the stinger) in the flesh of the intruder. The outcome depends on how well the sting is inserted, whether the stung animal (or the stinging insect) can remove the sting, the composition of the venom, and whether the target is susceptible to venom. The insect stinger is the original biological syringe, complete with a needle and a chamber that holds the liquid that's injected through the needle. Unlike a solid, tubular medical syringe needle, the insect sting shaft is composed of three parts, two of which slide in channels along the third immobile part. The sliding design overcomes the problem of the insect's small size. Imagine a mouse-sized doctor attempting to inject an antibiotic through a syringe into a patient. Could the doctor be large enough to grasp the syringe barrel and strong enough to push the needle into the flesh and depress the plunger? The insect's self-penetrating stinger solves these problems. The stinger works its way in through muscles that first slide one mobile side of the stinger deeper into the flesh and then the other side. Backward-facing barbs on the sliding stinger components help keep the embedded part from sliding back out while the moving side is inserted deeper. Instead of a thumb on the syringe plunger, insects solve the delivery problem in several different ways. For some stinging insects, a gizzard of muscles surrounding a sac of venom forcefully expels the venom. For other insects, a valve system inside the stinger assists in pumping venomous fluid through the hollow stinger shaft, usually aided by fluid pressure caused by muscle contractions, which telescope abdominal segments inward, generating the necessary pressure to squeeze the venom through the stinger and into the body of the target.

This marvelously functional device, the stinger, came from evolutionarily humble origins. Deep ancestors of stinging insects were sawflies, vegetarians that, despite the name "fly," are primitive wasps that use a stiff, hollow ovipositor to bore through plant tissues and stems to lay eggs in protected places. This tube was the key to the evolution of the stinger. The stinger is a hollow drilling tube that delivers venom instead of eggs into targets. A large series of developmental steps occurred between the ancestral sawfly ovipositor and today's ant, wasp, or bee stinger. A notable intermediate step is exhibited by parasitic wasps, which continue to use the ovipositor/stinger for depositing eggs but add venomous fluid that paralyzes or otherwise aids in preparing the host as suitable food for future offspring. Stings by parasitic wasps generally cause little or no pain in humans, an indication that parasitic wasp stings have not yet evolved a meaningful defensive role. The significant evolutionary change in the stinger, which dramatically altered its role, was the addition of a cocktail of venom components, concurrent with the elimination of the role as an ovipositor. Eggs were now delivered through an opening at the base of the stinger, freeing the stinger to function solely as a venom-delivery device.[3] Liberated of its egg-laying role, the now pure stinging apparatus was free to evolve venom that was active not only on hosts but also for defense against predators. Dual paralytic and defensive venom components continue to be seen today in many primitive ants and some solitary wasps, but this dual role is absent in all bees, an enormous group of 20,000 species that have lost the use of live animals for food, mainly replacing them with a vegetarian diet of pollen and flower nectar. In bees and social wasps, the role of the stinger and venom is strictly for defense against predators, with occasional use against other competing individuals, as witnessed during death fights among newly emerged honey bee queens and usurpations of established colonies by invading queen yellowjackets. Most advanced ant species primarily use venom for defense; however, on occasion, venom is used to capture prey.

"Careful, don't let him sting you" is an all-too-familiar phrase to warn against stinging insects. But male stinging insects do not sting. You read right. Males do not sting. Why not? The answer could not be simpler—they do not have a stinger! Even if a male bee (or ant or wasp, for that matter) attempted to sting, it lacks the equipment. The stinger is a highly derived, egg-laying tube, and males cannot lay eggs. They simply cannot evolve a stinger similar to a female's stinger. Consequently, males are harmless, have no ability to hurt large predators, and do not even aid their sisters to defend against predators. Threaten a male bee or a wasp, and it flees or hides. One of the more enjoyable aspects of teaching children about stinging insects is to reach into a jar and grab a robust buzzing honey bee. Invariably, the surrounding group gasps, in wide-eyed awe. How did I do that? Was it magic? Did I have a special mental power to control the bee? The bee was, of course, a male, often derogatively called a drone, and harmless. If only all such teaching were so didactic. In Arizona, the difference between male and female stinging insects can be vividly illustrated. Huge black carpenter bees, many times the size of a honey bee, abound in the spring. An audience's initial surprise at seeing me pick up one of these giant bees turns to shock when I gently place it between my lips. I don't mention that these wood-chewing bees have powerful jaws and can bite. No matter, the lesson is conveyed, although few are willing to volunteer to repeat what they have seen.

The previous example should not imply that males have no defensive tricks of their own. Nature has a never-ending store of surprises, illustrated magnificently by male bees and wasps. In place of a stinger, a male bee or wasp has hardened genitalia (the term for insect genitals) used for grasping the female during mating and transferring sperm. Male genitalia are a showcase of structural plasticity. Each species has minor to major structural differences from related species, which reduce the probability of mating between different species. This plasticity is also a preadaptation for evolving useful defensive structures, in this case, sharp, pointed sting-like projections from the end of the genitalia. When grabbed, males exhibit remarkably realistic stinging motions, and jab these hardened pseudo-stings into the skin of the captor. The predator's automatic response to being stung is to let go of the stinging creature. A pseudo-sting is usually sufficient to secure release, even from an experienced entomologist who intellectually knows better but who is overcome by natural instinct. To my chagrin, I have been tricked by a male wasp and lost a coveted specimen.

The sting is effective for only one reason: the venom. Venom is a liquid blend of materials that can be injected through the stinger. Most venoms consists of small water-soluble proteins, peptides, biogenic amines that also act as neurotransmitters within animal bodies, amino acids, fatty acids, sugars, salts, and a few miscellaneous compounds. Some insect venoms, notably from fire ants and their relatives, consist of alkaloids in chemical classes similar to coniine, the compound from water hemlock that Socrates was forced to drink. Other ant venoms are terpenes that smell like pine. All of these venoms act when they are injected below the protective epidermal barrier and into the body. Many are inactive if applied on the skin, because they cannot penetrate the skin to target vulnerable tissues and the bloodstream. The lack of penetrability of proteins, peptides, and biogenic amines, in particular, limits their use in traditional chemical defenses that are sprayed, dabbed, or oozed onto adversaries' skin. By enabling delivery of active components beneath the skin, the sting and venom opened a bonanza of opportunities for the evolution of highly specific and active components, particularly proteinaceous materials.

The sting and venom notwithstanding, life can be tough for a stinging insect. No defense is automatic. Just because an insect can sting does not mean that its sting is successful. Predators are not without defenses against stings. Notable first defenses against stings are dense, thick hairs covering most mammals; tightly overlapping layers of feathers on birds; hard, tough scales on reptiles; and rubbery, slippery skin on amphibians. These barriers can be difficult to overcome, especially for a single or a few individual stinging insects that must also navigate through the snapping, slapping, and scratching defenses of an aware and moving adversary. Often only tiny areas, typically around the eyes, nose, and lips and perhaps the underbelly of the attacker, are penetrable by the insect and its sting. The insect must recognize and then succeed in reaching these areas to achieve success.

Once the defensive barriers of an attacker are breached, other problems arise, such as delivering sufficient venom to inflict meaningful pain or damage that conveys the message to cease the attack. One advantage warm-blooded mammals and birds share is enhanced quickness relative to cold-blooded animals. Avian and mammalian quickness often means the stinging insect, with its recently planted stinger, is brushed off before much venom is delivered. Stinging insects possess two potential tricks in this ongoing evolutionary war game to help overcome the problem of slow or limited venom delivery. The first is to enhance the speed of venom delivery by enabling almost instantaneous delivery. This is achieved with powerful muscles surrounding the venom reservoir. The strength of these muscles can be seen in the stream of venom sometimes sprayed a distance of up to a foot through the air by social wasps. This delivery system ensures that a fair dose of venom is delivered before the insect can be removed. A second means to overcome the venom-delivery problem is called "sting autotomy." As the word suggests, in these species, including honey bees, several social wasps, and some *Pogonomyrmex* harvester ants, the stinger acts as a separate, semiautonomous unit from the rest of the insect, a unit retained in the skin of the victim by back-facing barbs and pulled out of the body of the insect as it retreats or is brushed off. The remaining small sting apparatus, unnoticed by the target animal, continues to pump venom from the reservoir via muscular action coordinated by a ganglion in the autotomized stinger. This system of autotomy ensures that complete venom delivery is achieved, thus maximizing the effectiveness of the sting.

Stinging insects and their venoms sometimes face two further defensive hurdles during the thick of battle. The specific species of predator encountered might not be affected by the venom, which might have evolved effectiveness mainly for another type of predator. Harvester ants are an example of this. The primary predators of harvester ants for which the venom evolved are vertebrates. In mice, harvester ant venom is the most lethally potent insect venom known; in contrast, it is less than one hundredth as lethal to insects. The difference in activities is related to the chemical composition of the venom and how it affects different animal physiologies. Another problem, equally problematic for the insect, is the predator's evolved resistance to the toxic effects of the venom. In this situation, the target physiology was originally vulnerable, but the animal evolved mechanisms to block the toxic action. Our old friend the *Pogonomyrmex* harvester ant is again a good example. Its main predators are horned lizards, which eat them with impunity. Why don't the stings, which would easily kill a mouse, affect the lizard? The answer lies in a venom-neutralizing factor in the blood of horned lizards. This factor renders horned lizards 1,300 times more resistant than mice.[4] How commonly stinging insects encounter this problem is scientific terra incognita.

Let's return to our hypothetical stinging insect. If she could speak, the first words she might shout to visitors are, "Who is at my door?" Her second words might be, "I sting."

THE FIRST STINGING INSECTS

One of the characteristics that sets man apart from all other animals is a need for knowledge for its own sake. . . . All knowledge, however small, however irrelevant to progress and well-being, is a part of the whole. —Vincent Dethier, *To Know a Fly*, 1962

Biology is the economics and energy of life. With only a limited amount of energy and raw materials to go around, life forms scrabble to get their piece. Essentially all energy for life comes from the sun's radiation. Only plants and other photosynthesizing life forms can capture this energy and turn it into useful molecules. (Oh, there are exceptions, including deep-sea thermophilic bacteria that live on chemicals released by deep-sea hydrothermal vents, but let's ignore them for this discussion.) Sunlight is a limiting resource that plants must compete for by growing taller, adapting to places where other plants cannot easily survive, or battling nearby plants with chemical warfare and other tricks. Materials in the form of carbon, nitrogen, oxygen, water, phosphorus, sulfur, potassium, magnesium, and a mind-numbing array of other elements needed for life are also in limited supply or in limited availability. Plants can make energy-rich and structurally necessary molecules from light and raw elemental materials, but they cannot make the raw materials themselves. No plant can make magnesium, for example, and thus plants must compete for these basic raw materials in addition to competing for light.

Life for animals also boils down to energy and materials. Animals, unable to photosynthesize with light, must obtain all their energy from either plants or other organisms that ultimately get their energy from plants. A small energy contribution from basking in sunlight to warm the body also occurs, a contribution dwarfed by the importance of photosynthesis. Animals are thereby forced to be herbivores (predators of plants), scavengers of former life, or predators of other animals, fungi, or microorganisms. Like plants, animals cannot make basic elements, including magnesium, and usually, but not always, need to obtain the basic materials of life from their food sources (macaws and elephants obtain some minerals by eating clay). Animals also cannot synthesize many essential molecules, including amino acids, vitamins, and some fats, and need to obtain these from their food sources. Overall, an animal's life is a continuous struggle for energy and materials in a world with millions of other species struggling for a similar set of limiting materials and energy.

Human societies use money as the basic economic unit, a resource approximating energy and materials. Money, though important, is not the total force that drives a human economy. Food, shelter, reproduction, and safety are the true drivers of societal life. Money is the currency for achieving these. The same drivers—food, shelter, reproduction, and safety—apply to animal life. Energy and materials are the "money" for achieving these in animals. Without energy, an animal cannot find food, cannot find or make shelter, cannot reproduce, cannot maintain a safety net, and cannot obtain materials. Animals are entrapped in a hamster-wheel world where energy is obtained from food, and food is obtained by using energy. A necessary requirement in this loosely circular world, as Gene Odum, my former ecology professor at the University of Georgia called it, is for the animal to obtain more energy and essential nutrients from its food than are required to locate, capture (if necessary), process, and digest the food. This requirement of obtaining more energy from food than required to obtain it was a key factor in the evolution of the insect sting.

Insects, though small and often dispersed in the environment, are nutrient-rich, dense packets of food, perfect for attracting the attention of hungry predators. Plants generally possess a much lower density of nutrients, contain much indigestible material, are harder to digest than insect or other animal materials, and usually contain nasty toxic compounds. Compared to plants, insects constitute an ideal source of food, but they are small compared with vertebrate animals. Size counts. In life's economy, a small bit of food is less valuable to a large predator that might expend more energy obtaining that food energy than it receives from consuming the food. Strange as it seems at first this cost-benefit relationship spares insects from having to defend vigorously against many huge predators. One simple, frequently used but sufficient defense consists of hiding in "plain sight" by cryptically resembling the resting background, a strategy that limits detection and increases the searching costs for the predator. Another commonly effective insect strategy is to flee rapidly and evasively, sometimes with a startling effect similar to humans' reaction to encountering an explosion of escaping quails. Flight gives a time advantage to the prey, and the confusion increases the cost to the predator attempting to capture the prey. Aposematic warnings and mimicry of aposematic insects are also often sufficient defenses. A disadvantage of advertising oneself is the possibility that detection might result in an attack; counterbalancing this attack risk is the inherent hesitancy of most predators to attack a potentially nasty prey and waste energy and time in the process of the failed attempt. In the classic example of Lincoln Brower's blue jay, the bird vomits after eating a monarch butterfly. Everybody knows the inordinate unpleasantness of stomachaches and vomiting. This misery is genetically hard-wired as nature's way of protecting animals against repeating the behavior that caused the vomiting. After its unpleasant encounter with a monarch butterfly, the blue jay refused future monarchs.[1] Not only did the bird endure the misery of discomfort and the loss of expended energy, it also suffered the additional insult of losing the hard-earned energy from the previously captured prey already in the stomach.

One small insect may be safe from large powerful predators because it is not worth the predators' effort. But what about a collection of small insects? Most people would not cross a room for one small blueberry, but if a bowl of blueberries is present, the story changes. Blueberries are now worth the effort. The same principle applies to a collection of insects. An aardvark is not likely to pursue a single termite, but it lives on collections of termites in termite colonies. Collections pose a serious problem for insects. A single insect's usual defenses are no longer as relevant. Better defenses are needed. Most termites nest underground where the soil barrier increases the difficulty and cost to the predator. In addition, termites might produce specialized soldier castes whose sole role is to defend against predators, large and small, with powerful, sometimes razor-sharp mandibles or by spraying sticky or turpentine-like compounds on their adversaries. These compounds are sprayed from the heads of soldiers, making these soldiers the original "nozzle-heads." Aggregations or collections of insects can thwart predators by confusion, as illustrated by simultaneous quail escape flights or whirligig beetle groups on a water surface. The predator becomes confused and cannot readily focus on a specific individual. A different but common defense by collections of insects is toxicity. In conjunction with their bright colors, ladybird beetles, commonly called ladybugs, contain toxic coccinelline and other compounds that taste bad and can sicken predators. Blister beetles, the source of the famed "Spanish fly," produce cantharidin, which is stored in their blood. Cantharidin is a general tissue irritant as well as potentially lethal toxin to humans, which achieves its reputation as an aphrodisiac by irritating the genital tract, thereby drawing attention to that area.

Ancestors of stinging insects likely lacked most of the above defenses of aggregations or collections of insects. As solitary insects, they experienced relaxed selection pressure from vertebrate predators. If any of these ancestral sawflies did aggregate, as is seen in some sawfly species today, they probably also had nasty chemical defenses, as do today's representatives. Sawfly dietary life was tough. They eat mainly pine needles and leaves from living trees or bore into fibrous plant stems. Nutrient levels in these materials are low, and toxins are usually high, but sawfly life gave them the preadaptation of a sawing, penetrating egg-laying ovipositor for boring into wood. Insect larvae in wood presented a much richer new food source for sawfly ancestors of stinging insects than the wood itself. Thus, a shift occurred from herbivore to predator, technically from herbivore to parasitoid. A parasitoid is an animal that, during its immature stage or stages of development, lives in or on the body of a single host individual, eventually killing that individual. Common examples are ichneumonid wasps that sting, sometimes paralyzing, caterpillars or other prey and lay eggs within the body of the prey. The eggs hatch into larvae that consume and ultimately kill the prey. Other examples of parasitoids are tachinid flies that lack stingers but accomplish the same goal by laying eggs on the host prey. The eggs hatch and burrow into the prey where they feed and develop. Ichneumon wasps, along with other parasitoid wasps, are examples of lineages that evolved from this sawfly-parasitoid ancestor. All of these solitary parasitoid wasps that use their sting-ovipositors for stinging prey and depositing eggs are subjected to little predatory pressure from large predators. They rarely sting entomologists who remove them with fingers from insect nets, a procedure not recommended for removal of honey bees or yellowjackets. On the rare occasions when a very large ichneumonid wasp actually manages to sting a person, the sting pain is typically trivial, confirming that defensive stinging against vertebrates is not a behavior that was developed, in part, because the sting and venom are ineffective and nearly useless defenses.

A major milestone in the evolution of stinging wasps, ants, and bees was a functional shift of the parasitoid wasps' sting-ovipositor to a dedicated stinger. This group with dedicated stingers is called the Aculeata. The name comes from the Latin word *aculeus,* meaning a stinger, and is a good description of the group. The significance of this modification is that eggs no longer needed to pass through the stinger, and the glandular secretions associated with the egg and its passage through the narrow stinger tube were free to evolve new functions, as painful and toxic defensive venoms. Like parasitoids, the original aculeate Hymenoptera were solitary, a lifestyle the majority of aculeates still maintain. These individuals offered little nutritional quantity to large predators. Consequently, they were not strongly targeted by vertebrate predators. Even today, most solitary aculeate Hymenoptera rarely attempt to sting in defense, and when they do sting people, their stings rarely hurt. Nevertheless, the wasp ancestors of modern wasps, ants, and bees occupied the pivotal position at the cusp of one of nature's greatest evolutionary achievements: the generation of the aculeate Hymenoptera, a group of some 100,000-strong species that changed and dominated the world.

With the innocent-appearing change of an ovipositor into a stinger, a major radiation of species was set and ready. All that was needed were behavioral and venom composition changes that enabled the nascent aculeate to expand its diet breadth by exploiting many available potential new hosts. Concomitant with expanded opportunities for new host food sources were problems of encountering new predators. Without some means to blunt the new suite of hungry predators, the ability to expand into new biological niches could not be realized. Here is where the stinger became crucial. With no role in laying eggs, and no important role in any of the insect's bodily functions, the sting was liberated to be molded into radically new roles. An important new role was the production of secretions that contained pain-inducing or toxic components. As populations, species numbers, and time and activity of stinging insects increase, notice by predators and attacks would increase. All that was needed in the evolution of a new defensive role for sting venom was for insects to sting the predator and for that sting to result in the venom bearer's escape. A sting that by chance mutation or genetic recombination caused even some pain would be more effective than a painless sting. The genes of even these few escapees would be passed along to future generations, initiating a cascading series of changes in venom chemistry, each more effective than the last.

As long as individual stinging insects remained solitary, the individuals presented too small a meal to attract attention of large predators.

Therefore, little pressure for the evolution of powerfully painful venomous components occurred. Aggregations of individuals often have advantages not available to solitary individuals. All individuals benefit if one individual finds a large food source and attracts others of the aggregation to share in that rich opportunity. This is biological tit-for-tat. You help me. I help you. Nobody loses, and we both benefit. Of course, this is not a conscious decision by participants who are unaware of the process. It is a process that favors individuals who, for whatever reason, happen to act in this fashion. Aggregations also benefit locating members of the opposite sex for mating. A cost tagging along with the benefits of aggregation is the greater nutritional reward available to large predators, who now find predatory efforts worthy. In this aggregated situation, a painful sting is beneficial and favored. The stage is set for an arms race: The prey evolves more painful and effective venom and stings, and the predator evolves new means of overcoming the defensive stings and venoms.

The ultimate form of aggregation is sociality. When a species becomes social, many individuals live together, usually in a protected nest, share in rearing offspring with overlap in generation with the adults (parents and adult offspring present in the nest at the same time), and different individuals specialize in performing tasks, such as egg laying, foraging, or defense. A serious disadvantage of social life is the burden of protecting the immobile, immature members from predators. Eggs, larvae, and pupae are highly sought for their excellent nutrition and digestibility, and they cannot flee. Instead of escaping, tending adults need to guard the home fort if they are to protect their young. Selection pressure from predators would disfavor social aggregations or nests of ill-defended prey, leading to the expectation that social insects would be rare. This is far from the case, given that social insects comprise an enormous portion of the animal biomass in most ecosystems.[2] What explains this paradox? The answer: the sting. The painful and toxic effectiveness of the sting was honed in parallel with the increased predatory pressures of vertebrates. Genetics and selection pressures at the group and population levels were the ultimate causes of the evolution of sociality. The venomous sting was a major, perhaps the most important, proximate cause enabling the evolution of higher sociality to proceed.

Among the small subfamily of slender, wasp-like ants, the Pseudomyrmecinae are species with two strikingly different lifestyles. One, represented by *Pseudomyrmex gracilis,* consists of small colonies of relatively large ants secreted deeply inside protective twigs and stems. They forage mainly for sugary foods and timidly retreat at the least threat. The other, the bullhorn acacia ants, *Pseudomyrmex nigrocinctus,* are small ants that form large dispersed colonies that live within the large, hollow thorns of the acacia tree. Bullhorn acacia ants live off extrafloral nectar (nectar secreted from areas not within flowers) and protein-rich Beltian bodies (named in honor of their discoverer, Thomas Belt) of the host tree. These ants strongly defend their home and food supply—the acacia tree—from all predators, competitors, and intruders that might harm the tree. A mutualism evolved in which the plant houses and feeds the ants and the ants protect the tree. One prediction based on the hypothesis that the sting of social wasps, ants, and bees evolved in response to predation pressure is that social insects, having much to lose, should have more painful stings than those with less to lose. *Pseudomyrmex* ants provide an ideal test. Both the large *gracilis* and the small acacia ant are closely related and are in the same taxonomic genus. Their main difference is their lifestyles: one has little to defend; the other has much to defend. The prediction is that, despite the great size difference between the two species, the stings of the smaller bullhorn acacia ant would hurt more than the stings of the large *gracilis* ants. Fortunately, I had the opportunity to test this hypothesis in the tropical deciduous dry forests of Guanacaste, Costa Rica, and in Florida. When I touched an acacia plant in Costa Rica, the ants immediately swarmed up my hand and arm, stinging along the way. It was not possible to get the ants off fast enough. And the stings hurt—and in massive numbers, stings really hurt. The rapid multitude of stings tremendously enhanced the pain. *Pseudomyrmex gracilis* would not even sting; instead, it treated my arm as a tree branch and ran to the far side of the arm attempting to hide. When I grabbed it, it stung me, but the sting was trivial. This difference in painfulness, despite a twofold difference in weight between the two ant species, was clear. The prediction was supported, albeit a bit painfully.

The economy of life produces astonishing outcomes. Among these is sting autotomy, the gruesome process in which a stinging insect self-eviscerates, leaving its stinger embedded in the target's flesh. This suicidal behavior troubled Charles Darwin as he formulated his theory of natural selection. He pondered how killing oneself could promote passing fitness via descendants to future generations. An insect's self-evisceration could provide strong evidence against his theory. Amazingly, even though Gregor Mendel's genetics, much less the modern concept of DNA, were unknown to Darwin, he came up with essentially the correct answer. By facilitating the reproduction of your close relatives, mainly nestmates, your lineage would be passed down via relatives, because of your selfless sacrifice. Sting autotomy maximizes the pain and damage of a sting, thereby aiding in the defense of the colony against large predators.

Life's economy promoted learning and decision-making in both stinging insect prey and in their vertebrate predators. If an important prey becomes too "spicy," a predator has two choices: (1) it can abandon predation of the prey and lose meals, or (2) it can learn to avoid the spicy stings and keep its meals. The latter choice would be the obvious favored choice. Intelligence is not simply an accident of evolution. It comes with some cost from increased numbers of brain neurons and energy consumption. Consequently, intelligence must provide some benefit. One benefit is that some form of intelligence is required for learning. Learning, in turn, is valuable for future encounters, or potential encounters, with predators and prey.

One mid-May morning, as I was sitting in front of my computer writing, I glanced out the window just west of the monitor and noticed an alert western kingbird sitting on a dead branch. The bird was actively looking to the right and then to the left. Beautifully clad with gray caps and golden breasts, kingbirds are master aerial acrobats, snatching flying insects from midair. The bird was sitting above the flight path of an Africanized honey bee colony located on a mesquite tree 10 feet to the south. Periodically, the kingbird would fly to the north and out of sight, quickly returning to its perch. Each time it raised its head and swallowed something that looked like a bee. How could it do that? Bees sting. Bee stings hurt. Asian and African bee-eaters, a colorful family of birds, catch bees in their long bills and then bash their abdomens, where the stinger is located, on a tree branch. The assumption is that this action "defangs" the bee so that it cannot sting and purges it of venom. Could there be more to the story? Humans are generalist feeders derived from a long lineage of entomophagous primates. We are model predators with our own generalist predator sense of taste representative of other general predators. To test the idea that there is more to honey bees than their stings, I captured a batch of bees from the flight path observed by the kingbird and froze them. (I did not want to be stung while eating them.) Next, a bee was divided into its three body parts: head, thorax, and abdomen. I chewed each part in turn to get a good sampling of flavor before spitting out the hard fragments of the body shell. Wow. Heads tasted like nasty, crunchy finger nail polish. The thoraxes were palatable, albeit the wings and legs added an unappealing plastic crunchy texture, and the abdomen flavor resembled a horrible combination of turpentine and a corrosive chemical. These flavors originate from exocrine glands in the worker honey bee. The head produces a mandibular gland pheromone composed of ketones, the thorax lacks large glands, and the abdomen contains the venom and the Nasonov gland that produces lemon oil compounds. No wonder predators might not choose worker honey bees: even if they don't sting you, honey bees taste terrible.

Back to the kingbird. Could it be deterred by the noisome taste of honey bees? If so, then what was it eating during these brief sorties in which something went down the hatch? As fortune has it, kingbirds, like owls, lack a grinding gizzard and regurgitate hard fragments from their meals. The fragment pellets are dropped below their roosts. These pellets can be soaked in water and microscopically analyzed to determine the diet composition. The dense swath of bunny-ears cactus (a particularly unpleasant cactus with thousands of nearly invisible, glochid, or prickly, spines) below the kingbird perch was removed and replaced with clear plastic sheeting. Sure enough, within several days, numerous pellets were deposited. The pellets contained head capsules of 147 male bees and not one from a worker (the two are readily distinguishable by the round shape and huge eyes of males and the guitar-pick-shaped, small-eyed heads of workers). Male honey bees cannot sting, lack large exocrine glands, and when eaten have a custard-like taste and texture with some added crunch. Overall, males are quite palatable. The kingbird solved the problem by learning to distinguish male from female bees while on the wing and only preyed on the males.

Learning and decision-making occurs not only in vertebrates generally recognized for their intelligence but also in prey of vertebrates. Honey bees are well known for their ability to learn to forage at the best times, at the most rewarding flowers, and most efficiently to obtain the nectar from the flower. Could they also learn risks posed by predators and make adaptive decisions based on that learning? To test this, mature honey bee hives were threatened but not harmed. Threats consisted of blowing my breath directly into the hive entrance. I had discovered that mammals' breath odor is the single strongest stimulus for bee defensive attacks. The hives themselves were never touched or otherwise threatened or harmed. After threatening the colony by breathing into its entrance, I stepped back 6 meters and swept up all attacking bees with an insect net. Each day for two weeks this process was repeated at the same time of day. Enormous numbers of bees attacked during the first two days. By the third day, many fewer attacked. On all subsequent days, very few bees attacked. The number of bees in the colonies was about the same at the end of the two weeks as on the first day of the experiment. The bees had learned that my predatory threats were harmless and did not deserve a strong defense. Similar learning is routinely observed in South Asia where giant honey bees, *Apis dorsata,* commonly construct their hives above entrances to religious temples. Although people regularly enter and leave the temples with their heads passing only a few feet from the combs, the bees do not attack. These giant honey bees have also learned to evaluate risks posed by predators. I have never read of anybody reaching up to touch a comb but suspect that would be a most unwise decision.

4

THE PAIN TRUTH

To one accustomed to the flora and fauna of the forests of temperate regions, the forms, colors and odors of the tropical forests are a continual surprise. Fantastic forms and wonders everywhere meet the eye, until presently one becomes like a child wandering in fairyland, accepting seemingly impossible things as a matter of course. —Phil Rau, *Jungle Bees and Wasps of Barro Colorado Island*, 1933

Everybody knows what pain is. It is the sensation felt when a knee is scraped in a fall, the skin is overexposed to the sun, or a bare foot steps on a bee. Pain is familiar, yet mysterious. We know pain when we sense it. Pain is clearly recognizable. Warmth is not pain, though it can become pain if too much heat is experienced. Likewise, chills caused by cold temperatures are clearly unpleasant, but they are not pain. Cold, like warmth, can produce elements of pain but is not classic pain. As the tobogganing kid knows, toes get wet, cold, and unpleasant but not usually truly painful like a stubbed toe. But, oh wait until the true pain comes as the toes warm before a toasty fireplace. Though we might call stomach upset and nausea painful, is it really pain? We call nausea "pain" for lack of a proper descriptive word, but everybody knows nausea pain is different and, I would argue, far more unpleasant than the pain of a runner's stitch caused by the spleen contracting to squeeze more red blood cells into the bloodstream to provide acutely needed oxygen to muscles.

We know pain when we feel it. Do we know physiologically and medically what pain really is? The response gets murky. Describing actions that cause pain—slamming one's finger in a door, for example—is simple. Distinguishing pain from nonpain is generally easy. Hunger, though often called pain, is certainly different from the pain of a stomach ulcer caused as acids eat into and destroy stomach tissues. Again, we lack a common, distinguishing word for hunger pain, though perhaps "hunger pangs" is a proper phrase that when spoken sounds the same as "hunger pains."

There may be no universal consensus on what is and what is not pain. We generally recognize pain as a distinct experience, one that presents in a variety of flavors. Common pain is the sensation felt when skin is damaged, a tooth injured, a bone broken, a muscle pulled, the spleen makes a stitch, or a variety of other mostly dermal or skeletal-muscular problems. Another broad category, visceral pain, is experienced when visceral organs signal damage or potential damage. The visceral pain resultant from tonsillectomy in adults, hemorrhoid surgery, childbirth (I am told), or other sources one hopes to rarely experience is distinctly different from common pain and a considerably less pleasant form of pain. Headache pain is another category of pain. The point of this discussion is not to define pain concretely, to make a pain phylogenetic "family" tree, or even to claim that all the separations above are clear-cut (they are not); the point is to illustrate just how complicated and murky pain is.

How is the nebulous and murky nature of pain explained? What is the cause of this lack of tidiness, both in descriptive language and in real-life sensation? Several poorly defined explanations can be offered. A medical explanation might focus on the separate pathways of nerves and distal structures between the motor and autonomic nervous systems and the sensory nervous system. Action potentials sent from the brain to muscles travel down different nerve axon pathways than pain signals generated when the tongue is bitten. Signals to the brain originate from receptors located throughout the body. Many of these receptors are sensory receptors that detect temperature, pressure, stretch, chemicals, itch, or a variety of other sensations, including pain. The signals from these receptors are transmitted to higher nervous centers in the spinal cord and brain through fine nerves of the separate sensory nervous system. Matters become more subtle. Pain and itch, for example, are separate sensations.[1] Are they related, that is, is one just a smaller degree than the other? No, they are not simply degrees of difference, and, unfortunately, how they relate is unclear and an active topic of research. Is the tickle sensation related to the pain and/or itch sensation? Again no tidy answer is forthcoming. Complicating the issue further is the feature that tickling can be a pleasant sensation, especially in social situations, or it can be an excruciatingly unpleasant experience. How are the two tickle responses related? In degrees of stimulus? The difference is unclear.

Is pain always unpleasant or, as pointed out by my grade 7 science teacher, can it also be pleasurable? A love-hate situation occurs when baby teeth are about to fall out and to be replaced by permanent teeth. The loose tooth hurts, but the urge to wiggle it is irrepressible. We can wiggle the tooth just enough to cause a little pain, an enjoyable sort of pain, but not too much. And we can control this dynamic precisely and entirely ourselves. What is the difference between the two tooth pains? Is it simply strength of the nervous signal emanating from the tooth receptors? Probably not. Here, other important players in the pain system come into play, the higher processing centers of the spinal cord and brain. These centers filter and process the signals to determine the importance of the signals and then send them to our conscious centers of the brain. If the signal indicates a dire situation, as in a hand placed on a hot stove burner, the processing centers convey outside the conscious pathways to the action centers to signal reflexive removal of the hand. The conscious centers are involved in the process of learning for directing future behavior to avoid placing the hand on a hot object.

Pain serves a higher purpose in the biology of life than revealed by analyses of nerve pathways and processing centers, however interesting these might be. Why should pain, a most universal sensation in living animals, exist at all in nature? Certainly not for pleasure or torture. Only adaptations and sensations of value for promoting the life, survival, and reproduction of an organism stand the test of time.

Pain is a basic sensation of life, experienced by all animals. Even the simple single-celled paramecium moves away when it encounters the high acidity from the drop of vinegar placed in its watery bath, just as we jerk our fingers from a hot stove. Does the paramecium experience pain? Certainly not in the way humans do, as it has no brain or self-awareness, but it responds the same as we do to the negative situation, so in practical terms, we can call it a pain response. In biological terms, pain is simply the body's warning system that *damage has occurred, is occurring,* or *is about to occur*. Nothing more. Pain is not damage. It is merely a harbinger of damage. Is pain truth? Perhaps. If damage occurs concurrent with pain, then pain is truthful in sending the honest signal that the body is at risk and has been compromised. A bruised shinbone sends truthful pain.

What if pain is intense and no meaningful damage occurred? Is that pain truthful? This paradox of the veracity of pain's role, *damage is about to occur,* is just what stinging insects exploit. Returning to the bee and the foot, the sting to the sole elicits pain and lifting the foot is a response that benefits the bee (well, maybe no longer to *that* bee, but to her nestmates). Has meaningful physical damage to the person been done by this sting? Often, the answer is no. Stinging insects are masters at exploiting this weakness to their benefit in the honesty of the pain signal. To stinging insects, we might simply be fools who fall for the trick. To us, it is better to be safe than sure; thus, we believe the signal is true. If the damage were real, the downside cost could far outweigh any benefit obtained by ignoring the pain. Why take a risk? In life's risk-benefit equation, the risk often dwarfs any potential benefit. Herein lies the psychology of pain. Unless the animal or human can know that a rainbow of benefit is awaiting on the far side of the pain, natural psychology dictates not to chase the rainbow.

Pain can be a lie. The insect sting exploits a weakness of the pain signal system to propagate a masterful deception. That deception, the lie, benefits the stinging insect by cheating its adversary out of a meal, perhaps cheating it out of the use of space near the insect or its nest, or even cheating it out of some other resource, such as a feeding site.

In most painful circumstances, it is adaptive for the stung individual to accept the lie and ensure its safety. One can lose many small meals and survive. One serious poisoning and the individual might not survive. The math is on the side of caution.

For every liar and cheater, someone out there is not fooled. For the lying pain of stinging insects, some animals and people have broken the trick: they ignore the pain and reap the rewards offered by the stinging insect. In much of North America, skunks are common denizens of the rural countryside. Beautifully adorned in tuxedo black with brilliant white stripes or spots, skunks are known mainly for their aromatic properties, but they are also efficient predators of insects and other small game. Skunks have a fondness for stinging insects and avidly dig out and consume the contents of yellowjacket wasp nests. They also enjoy honey bees, another spicy dietary staple, and have learned to discount the pain. Bears are another trick-breaker, famously known for their love of honey. Bears tear apart beehives in hollow trees or beekeepers' boxes, relishing the sweet honey and rich brood, all with apparent impunity to the bee stings. Common wisdom dictates that the dense fur of bears protects them from stings, but this half-truth wisdom mainly protects our empathy for the bear and its potential pain. In real life, the bear suffers many stings, especially around its sensitive eyes, nose, ears, tongue, lips, and mouth. It has learned that a certain number of bee stings can be endured without injury and that the reward is worth the pain. Likewise, for the fabled African ratel, or honey badger. Ratels, relatives of wolverines, are medium-sized black-and-white, tough-skinned intrepid animals that routinely feed on all sorts of prey, including poisonous snakes (reputed to be unable to bite through the tough skin), chase lions and other carnivores from their prey to claim the kill, *and* are best known for their love of honey and bee brood. Ratels, like bears, learned that a certain number of stings cause no meaningful damage, and thereby they have learned to overcome the pain. This is a tricky game for ratels. Bee stings are truthful as well as painful. Enough stings, about 4 for a mouse, or an estimated 140 stings for a ratel, can kill. Until around a hundred stings, the ratel is safe. No one knows how well ratels can count in the ratel–bee brinkmanship game, but they likely can sense when a dangerous level of envenomation is near. The game can be tricky, however, for some ratels have misjudged and paid the ultimate price of being stung to death.[2]

Truth, like beauty, can be in the eye of the beholder. Pain truth comes in two flavors, imagined and realized. With stings, our imagination is vivid and strong, even if the sting pain is not realized. The paper wasp, *Polistes instabilis,* provides a real-life example of pain. Perhaps the name *instabilis* tells us something. No matter the actual origin of the name, their behavior appears unstable to people walking through the scrubby brush of their tropical habitat. Their presence is usually painfully detected as a sting to the back of the neck or bare arm subsequent to brushing past a leafy tangle with a nest attached to a small branch. This truthful pain was realized by our cowboy guides, as they led a group of biologists on an expedition through the thickets to the most northerly location of the magnificently beautiful military macaws. After the second in line was stung on his wrist and yelled, we all stopped to allow the wasps to return to their nest and become calm, albeit alert. Up to this point, our guides had considered us a bunch of inept, cowardly biologists. We needed to both move forward and change the guides' perceptions of us. As the only entomologist in the group, I obviously needed to take charge. Here's where knowledge of stinging insects is crucial. For wasps and bees, the two greatest factors that stimulate attack are human breath and rapid movement. To continue our trek, both factors had to be minimized. Theory is fine, but reality was calling. I carry a 2-liter, wide-mouthed plastic jar for rare opportunities, and this was the perfect occasion. With eyes glued on every movement and hint from the wasps, I held my breath and slowly advanced, jar in my left hand and lid in my right. During this eternity of 30 seconds, the jar was snuggled underneath the nest and the lid just above. Snap. The lid was on the jar and all wasps inside. Except the branch prevented the closing of the lid. A yell for assistance brought a cowboy with a machete to cut off the twig, securing all wasps inside.

This reverse showmanship worked: rather than showmanship on the part of the wasps, it was showmanship by the predator, which fooled the wasps *and* earned biologists some status.

The Australian bull ants, sometimes called bulldog ants, are inch-long, lithe creatures with enormous eyes, long mandibles, and lightning speed. And they jump. Their uncanny behavior of turning their heads to follow observers adds to their mystique. In Australia, they are highly respected, if not outright feared, for their fabled stinging ability. Among all of Australia's native insects, bull ants head the list of painful stingers. This is partly because Australia has no native honey bees, no hornets, no yellowjacket wasps, and their social wasps are mostly in the generally placid genus *Ropalidia*, a group similar to many *Polistes* paper wasps in Europe and the Americas, though generally milder in disposition. Hence, Australians lack comparisons between their bull ants and other painful stinging insects around the world. Given the background of stories about bull ants, I approached collecting them with some anxiety and caution. However, I didn't know about their athletic abilities, something mentioned but frequently glossed over by writers of articles on these ants. As I collected some individuals from a nest, an alarm was sent and a boiling mass of ants issued from the colony. My athleticism didn't match theirs, and the feared stings were realized. I was stunned, not by the pain, but by the low level of pain. The balloon of anticipation had been deflated. Why did the stings not hurt so much? The pain was less than the sting of a honey bee. Flare and swelling were also minimal, and the pain was short-lived. Had I been stung too many times and simply could no longer detect pain? This was a valid concern, so how could I address it?

As fortune would have it, the greatest congress of social insect scientists was meeting about that time in South Australia. Midway through the meeting we took a break, climbed on some buses, and visited Kangaroo Island. On the return trip, the driver spotted a large bull ant colony along the side of the road and asked whether we would like to stop. A resounding yes echoed throughout the bus. Ah, opportunity. My reputation with insect stings was well established. That set the stage for quiet showmanship and trickery to demonstrate my ability to evaluate pain. Normally, it is unfair to expose people to insect stings, but this group of experienced social insect colleagues was fair game. I went to a colony, picked up individual ants, and dropped them in a jar. Others saw this and realized my approach was much faster and easier than trying to pick up hypermotile ants with clumsy forceps. Sure enough, five colleagues got stung. I ask them casually, "Does it hurt much? How does it compare to a honey bee sting [all had been stung by honey bees]?" In all five cases, the reply was that the sting was surprisingly less painful than expected and hurt less than a honey bee sting. Apparently, my sting pain detection system operates well.

Truth, lying, and cheating are not solely in the domain of female stinging insects—the ones that can sting—and in humans who exploit or study them. Males of some stinging insects can lie about stings and pain also. Although they have no stinger, no venom, and cannot harm a large predator, males can put on a good show. Because imagined sting pain is real in people and in other animals, male bees and even mimicking flies buzz ferociously when they are captured, as do female bees. The higher-pitched buzzing of a captured insect is an aposematic warning signal that conveys danger. This male auto-mimicry of stinging females is energetically costly and would be very unlikely to have evolved if it were not effective. The sharp spines on male reproductive genitalia, particularly in wasps, have a dual role of matching the structure of the female reproductive system and for providing a modicum of protection against large predators. Like many evolutionary questions, which was a more important selection factor—mate matching or defense—is unclear. Probably both factors were important. In addition to possessing hard, sharp spines, these males exhibit uncanny stinging movements in their similarity to the movements of females. When grabbed, these males curve their abdomens and jab the sharp spines into the fingers or mouth of the offender. Many an experienced entomologist, including me, have been tricked by this maneuver, and our instinct caused the release and escape of the male, to our chagrin. Score one point for the male wasp, zero for the entomologist.

5

STING SCIENCE

In physical science the first essential step in the direction of learning any subject is to find principles of numerical reckoning and practicable methods for measuring some quality connected with it. I often say that when you can measure what you are speaking about, and express it in numbers, you know something about it; but when you cannot measure it, when you cannot express it in numbers, your knowledge is of a meagre and unsatisfactory kind; it may be the beginning of knowledge, but you have scarcely in your thoughts advanced to the state of Science, whatever the matter may be.

—Lord Kelvin, *Popular Lectures and Addresses, 1891–1894*

Science is rarely sterile. Scientists are adventurers like ancient explorers sailing to undiscovered parts of the globe, who do not know what they will find or discover but seek the thrill of the unknown. Contrary to movie caricatures, scientists are not eccentric, crazy, brilliant people in strange laboratories concocting various magical brews or wild computer programs. Scientists are people, equally exciting or boring, like our usual acquaintances. Science is the process of discovery, distinguishable from other human endeavors. The discovery process is self-correcting; that is, if evidence disproves a scientific concept, that old idea is either discarded or modified consistent with the new factual information. In practice, this process is not usually as smooth or as rapid as described. Most scientists make their greatest discoveries early in their careers and, because they are human, become attached to their discoveries. Within the scientific community, new ideas stimulate new experiments to test the ideas, generating new facts and information. Good scientists will look at new facts and modify or outright discard their ideas if they are shown to be wrong. But this is difficult. Nobody wants to think that much of what he or she accomplished in life is wrong. Young scientists are typically spared emotional attachment to earlier ideas and form their ideas mainly based on current facts. Thus, science tends to progress through younger people, and old ideas tend to die with the originators of those ideas. Through this cynical view, science progresses one coffin at a time.

However imperfect science might be in its practice, it is the best system of discovery of the real world we have. Religion, based on fundamental, often ancient, immutable truths, which can change only painfully slowly and not strongly in response to facts, differs at the most basic level from science. Likewise, science differs from a variety of political systems based mainly on power, authority, and human personality. Science has the uncanny way of progressing, in spite of individual human or institutional personalities and other obstacles, toward ever better understanding of the world and the universe around us. Science is an exploratory process more so than a goal to realize. Yes, there are goals, and these must be clearly defined for funding agencies to support the research, but the real excitement and driving force in science is the adventure of seeking the goal, not in attaining the goal. Attaining the goal is, naturally, exciting for the pride, fame, and satisfaction it brings, but often the opportunities for more funding, talented collaborations, and the ability to continue expeditions into the unknown are more exciting than achieving the goal.

Ever since I can remember, I was fascinated with ideas and facts. At four years of age, the idea that 10 + 10 = 20 was fascinating. It was a fact: I could count 10 pennies, count into a new pile another separate 10 pennies, then mix the piles and count again. I got 20 pennies. When I was a little older, I remember reading Jean Henri Fabre, the great entomological observer, experimenter, and writer of the late nineteenth and early twentieth centuries. When he was about five years of age, he asked the question, "How do I see?" A simple, but profound question. We take for granted that we see through our eyes, but how many of us ever test that fact? The youthful Fabre designed a scientific test to determine how we see. He closed his eyes and opened his mouth. He couldn't see. Then he closed his mouth and opened his eyes. He could see. He concluded that we see with our eyes, not our mouths and experimentally demonstrated the fact that we see with our eyes. As trivial as this test was, it hooked me (and apparently Fabre) on the method.

A kid growing up in rural Pennsylvania has limited opportunities in some respects and enhanced opportunities in others. We had no professional sports teams, no elaborate amusement parks, not many shopping opportunities—the best being a row of toys at the five-and-dime store in the nearest big town—and few organized entertainments nearby. We did have trees, brooks, old abandoned fields, nice pleasant summers, and lots of insects. For unknown reasons, I was not attracted to dinosaurs or other large animals but found tiny insects fascinating, perhaps because they were small like me. They also opened a whole world, just to me, a world not appreciated by the other kids in the neighborhood. Particularly fascinating were the brightly colored paper wasps, yellowjackets, and assorted other solitary stinging wasps and bees. Honey bees were drab brown and not so exciting. The main exciting aspect of bees was their ability to sting. Butterflies, especially tiger swallowtails, were also fascinating because they were big, beautiful, and hard to catch. My parents endured, perhaps encouraged, my interests in nature.

School became progressively more interesting. First, math. Math was a simple, crisp subject so logical and challenging. Then came biology. Oh, my poor teacher when I fell into a swamp coming out all muddy and smelly after a failed attempt to catch a green darner dragonfly. The next year came physics, a subject radically different in material from biology but fascinating for its own beauty. The following year brought chemistry, my newest love and adventure. Chemistry brought unlimited experimental opportunities, not all appreciated by everyone, for example, the time when my experimental smoke "bomb" produced an enormous mushroom cloud of black smoke. College was next. With chemistry fresh in mind, it became the chosen subject. After six years of chemistry, including a move to the Pacific Northwest for a master's degree, I found chemistry lab work challenging but lacking. Chemistry lacked living, moving nature—insects to be exact. Stinging insects were still etched in my memories. Armed with renewed memories and enthusiasm, a move to Georgia was in order.

At the University of Georgia, I found myself among a group of bright students, all of whom had graduated from biology or zoology programs and were well ahead of me in many aspects of entomology. As an undergraduate, I was exempted from the biology requirement and did not take a single course in biology. In graduate school, I was among students and faculty who called insects by their scientific names. I knew the common names of the insects but not any scientific names. When it came to picking dissertation research, a natural was to combine what I knew best, chemistry, with what I loved, stinging insects. My professor, Murray Blum, wisely suggested I work on *Pogonomyrmex* harvester ants, locally available and nasty stinging insects whose venom chemistry was unknown.

With that goal in mind, off I went with Debbie, a talented zoology student who fortuitously happened to be my wife. We piled buckets in the car's trunk and, with shovels in hand, headed on an expedition to find harvester ants. The procedure was simple enough: find the ants; dig up the colony; put ants, dirt and all, into a bucket; and bring them back to the lab to study. Digging in the sandy soil of Georgia was a delight, not like digging in the limestone-laden rocky soils of the rural Pennsylvania Appalachian Mountains. We quickly got casual and relaxed at the easy work in an idyllic setting. Wham, an ant stung me. Serendipity had struck. This was no ordinary sting. This sting really hurt. The pain, delayed at first, became piercing and excruciating. Then, it progressed into waves of deep throbbing visceral pain, as Debbie, who also got stung in the operation, described as "deep ripping and tearing pain, as if someone were reaching below the skin and ripping muscles and tendons; except the ripping continued with each crescendo of pain." The pain was not at all like the hot burning pain produced by all my childhood experiences with stinging insects. All the stings of my childhood world of honey bees, yellowjackets, baldfaced hornets, bumble bees, and paper wasps resembled the pain of a burning match head that flipped off its stick and landed on my arm. All of these immediate, intense pains lasted only 5 minutes or less before receding to a tolerable, if not benign, level. Harvester ant stings were different. Not only was the sensation less burning, it lasted . . . and lasted. Four hours later, we were still in pain, albeit decreasing pain. After 8 hours, the last vestiges of pain were finally gone. From a chemist's and biologist's viewpoint, even more interesting were the other reactions. Harvester ant stings caused the hairs around the sting site to stand up on end, much like the bristling shoulder hairs on a frightened dog. There was no fright though. Something independent of the brain made these hairs stand up. Also, the area around the sting became moist with sweat, again something independent of the brain. No other insect stings we, or others, had experienced ever caused either of those reactions.

Thus, an interest in stings, their chemistry, biochemistry, physiology, and their biological roles in the lives of the insects and their targets was born. Two immediate questions came to mind. First, do all harvester ant species cause the same types of reactions? Second, do any other stinging insects cause similar reactions? These were untested ideas. Ideas are great but are not meaningful if no data exist to test them. Data were needed. Off we set on an entomological adventure to the western United States in search of data. With shovels, insect nets, maps, containers, a portable microscope, reference books, and ice chests all crammed into our old VW camper bus, we were on our way. The primary goals were to collect as many of the 20-some-odd species known at the time in the United States as possible, to collect their venom for investigation back in the lab, and to bring back live colonies. Our indirect goals were to compare the sting painfulness and reactions of the species. We had no desire to be intentionally stung, but if we did get stung, we might as well be prepared to record the data. Wasting a good opportunity for a data point seemed crazy.

Rumor had it that the Georgia species of harvester ant had a mellow, hospitable, Southern disposition compared with some wild western species, so we braced for what might come. Northern Louisiana is the eastern limit of the Comanche harvester ant, a generally unaggressive species that curiously sting-autotomizes; that is, it leaves its sting in human flesh, as do honey bees. The sting also hurts, like the sting of the Florida harvester ants in Georgia, only the pain lasts longer. In Texas, and fitting its proclaimed reputation for big, we ran into a larger harvester ant. This species, *Pogonomyrmex barbatus*, called the Texas agricultural ant by H. C. McCook, the famous popularizer of nature around the late 1880s, is also called the "red harvester ant," an essentially meaningless name because, with a few exceptions, all harvester ants are red. This species builds impressive nests, with an entrance hole in the middle of large circles of barren earth. These formicid engineering experts clear and maintain the bare areas. Their size and color belie their true prowess as purveyors of pain. Though not fakes, their stings hurt less than those of the Maricopa harvester ant, a smaller and more delicate species. Their stings not only hurt less, the pain lasts a shorter time, and they do not sting-autotomize.

In the small and delightfully charming town of Willcox in southeastern Arizona, we found Maricopa harvester ants. These ants were the most impressive species of harvesters of the trip. They dominated low-stabilized sand dunes around the Willcox Playa, a usually dry lake in a small basin with no outflowing stream or river, a mini Great Salt Lake Basin, so to speak, only lacking the salt. Perhaps because of the high water table near the lake, the Maricopa harvester ants build enormous mounded ant castles for the 20,000-plus ants in the colony. Except during termite swarming times, these Willcox ants are placid collectors of seeds and generally do not readily sting. Swarming termites can be viewed as mobile "seeds," packed with protein and fat and much easier to eat than hard dry seeds. The ants dramatically shift behavior during this time and become avid predators. Sandal wearing is not recommended. Don't let the delicate, lithe body shape or unassuming demeanor of Maricopa harvester ants fool you. The stings of these ants really hurt. The throbbing pain can last 8 hours, decreases only slowly, and the ants readily autotomize their stings into humans or other unfortunate animals. These ants were the most painful stingers we encountered on the summer's trip. To add veracity to their message, the venom of the ants at this particular location is the most toxic known ant, wasp, or bee venom, some 25 times more toxic than honey bee venom and 35 times greater than western diamondback rattlesnake venom.

Why do defensive stings of venomous insects hurt so much? Why should some defensive stings be toxic, much less highly toxic? After all, isn't the insect served simply by making the attacker release the insect and abort the attack? A first hint that points to an answer is that some insect venoms *are* highly toxic. High toxicity evolved independently many times in ants, wasps, and bees. Repetitive evolution of a similar property, especially when the molecules responsible for that property are different, indicates some function, not just random "mistakes" of nature. What possible function could venom toxicity have? The question becomes especially poignant when considering that the toxicity of harvester ant venom is 800 times more toxic to some predators than to an insect prey. The solution to this conundrum is revealed in the words "truth in advertising." Pain is an advertisement that damage has occurred, is occurring, or is about to occur. Without enforcement, advertisement becomes a slick system of lies. Intelligent animals can see through lies, or learn to see lies as what they are, and the advertisement loses meaning. In the insect sting system, pain is the advertisement, and toxicity is the truth. Toxicity is truth because it is real damage or death. The toxicity truth becomes especially important for small vertebrate predators that are more susceptible to damage than large predators. Without toxicity, a smart predator learns the dishonesty of sting pain and can ignore the signal. When this occurs, for example, beekeepers learning that dozens of stings pose no real physical threat, continue to rob beehives; the stinging insects lose. In the case of a 20-gram shrew or mouse, in which four honey bee stings can be lethal, the message of sting damage rings crystal clear. Thus, a gradient of effectiveness of sting toxicity spreads across the predator field, helping stinging insects survive attacks by some predators more than others, overall providing a net benefit in the game of life. Even predators as large as a 50-kilogram (110-pound) beekeeper are at risk of death from 1,000 honey bee stings.[1] Combined damage and lethality are crucial to the long-term evolutionary effectiveness of insect stings against intelligent predators.

The venom constituents that cause pain and those that are damaging or lethal are not necessarily the same. Selection pressure for pain came first. This we surmise based on its immediacy as a defense and the presence of painful stings in present-day wasps closely related taxonomically to the stinging ancestors. Examples of painfully stinging species whose venoms are essentially not toxic include some large ichneumon wasps (*Megarhyssa*) that are parasitic on wood-boring sawfly larvae, bethylid wasps, solitary parasitoids (insects whose young are parasites that eventually kill their hosts) of beetle larvae, velvet ants, solitary parasitoids of bees and wasps, and spider wasps, solitary parasitoids of spiders. The chemical nature of the painful venom components of these wasps with simple life histories is mostly guesswork, but likely, each type of wasp has a different chemical or set of chemicals responsible for the pain. The pain-inducing components for velvet ants include at least the biogenic amine serotonin (5-hydroxytryptamine), a known algogen when injected beneath the skin. Serotonin is also a pain-inducing component in a wide variety of social wasp venoms. Histamine, another biogenic amine, is widely present in venoms of yellowjacket wasps, paper wasps, hornets, honey bees, and some ants. Histamine primarily causes vasodilation of blood vessels resulting in swelling, warmth, redness, and some itching. It does not cause sharp pain. In that regard, histamine is not a strong agent of pain. Acetylcholine, a third biogenic amine, does cause sharp pain and is found only in hornets. These small molecules are not the important direct inducers of pain in insect venoms; that role falls to a variety of small peptides whose structures vary strikingly from one group to another. In honey bees, the painful component is melittin, a 26 amino acid peptide containing five basic amino acids. In wasps, the kinins, 9 to 18 amino acid peptides that cause heart pain among other activities, cause the intense burning pain. Harvester ant venoms contain barbatolysin, a 34 amino acid peptide that appears to cause pain. The pain-inducing agents in the various ant venoms are not known, though some species contain their own ant versions of kinins that likely cause pain and a wide variety of other peptides.[2]

Damage-causing venom components evolved subsequent to the early pain-producing agents. In most venoms, the identities of the actual toxic components are not known. In the best-studied insect venom, that of the honey bee, the most toxic component is the enzyme phospholipase A_2, which causes no skin pain. The more abundant but less lethal component, melittin, is the other important toxin. Melittin is a heart poison that also causes hot, burning pain by destroying cell membranes, including those of nerves. We recently isolated and are characterizing a lethal component from harvester ant venom that induces all the skin and pain reactions of a sting.

The Pain Scale

After we returned from the trip west with a car burdened with buckets of ants, urgent immediate and long-term questions arose. The immediate, and less-interesting, question was what to do with the ants. This was a less interesting question because the answer lay in dissecting and collecting massive quantities of venom for drying and freezing for future work. Massive quantities meant 5 milligrams or more, about 1⁄1000 of a teaspoon, of each type of harvester ant collected. At 40 ants needed to yield 1 milligram of venom, and at about 3 minutes to dissect each ant, gold is cheaper than harvester ant venom.

Long-term questions converged on determining the value of the venoms to the stinging insects themselves. The venoms evolved for the benefit of the insects, not for humans. What were these benefits, and how did they change the lives and biologies of the insects? To help answer these questions, the properties of the stings and venoms needed to be evaluated. Pain and toxicity are the two basic properties of each sting. To test hypotheses regarding pain and toxicity, each venom needed to be compared with the venom of other stinging insects. Then the lives of each species were compared to determine whether venom properties correlate with life histories. For toxicity comparisons, a variety of physiological and toxicological methods are available, each yielding a numerical value that can be compared to the numerical value of other venoms. In principle, toxicity comparisons are simple. But what about pain comparisons? There were no physiological or pharmacological methods to place accurate values on pain. Even today we lack reliable methods to insert electrodes into nerves or parts of the brain to measure pain, and then to understand the meaning of the electrode recordings. Likewise, interpreting the results of more advanced brain-scanning techniques vis-à-vis pain is unclear, though great progress is being made. Someday, we hope to measure pain quantitatively and cheaply. In the meantime, how could sting pain be measured numerically? What was the solution?

A pain scale was needed. The answer was simple; however, making the scale was not. A useful scale had to be reliable, reproducible, and indexable. There was precedent for a pain scale, albeit one designed for measuring human chronic pain. The McGill Pain Questionnaire, developed by Ron Melzack at McGill University primarily to measure chronic pain in patients, consisted of rating pain levels derived from patient questionnaires and caregiver evaluations of facial and body language.

Insect stings cause short-term, mainly ephemeral pain and a variety of nuances that relate to the person stung and the stinging insect. Sting pain induced by a single sting can vary depending on how much venom the sting delivered, where on the body the sting occurred (for example, stings to the nose, lips, or palms of hands hurt considerably more than stings to lower legs, arms, or top of the skull), the age of the insect, the time of day the sting was received, and other factors, including the sensitivity to pain of the individual. For reasons of consistency and reliability between different evaluations under different circumstances, only a few numbers were used in the sting pain scale. The scale ranges from values of 1 to 4 and is anchored by the value of a single honey bee sting (*Apis mellifera*), defined as pain level 2. The honey bee is a convenient reference point because honey bees exist nearly worldwide, are abundant, most people have been stung by a honey bee, and they are about midway within the range of pain intensities produced by wasp, ant, and bee stings. Also included on the scale is a trivial value of 0 for stinging insects incapable of penetrating human skin but possibly able to sting other animals. The criteria distinguishing between pain levels are that the pain of the lower level is substantially less than the pain in the upper level and that the evaluating person would clearly know that one sting hurt more than the other. When comparing species, the evaluator compares the current sting pain with memory of the pain of a previous sting by a honey bee or other species for which the pain was rated previously. In some cases, values halfway between whole numbers are assigned in which the pain appears distinctly greater than the lower level, yet distinctly less than the higher level. This evaluation system works remarkably well as witnessed by nearly identical ratings for stings by various colleagues. Chris Starr, a fellow graduate student colleague at the University of Georgia, and I spent innumerable hours discussing the topic. We also discussed ideas widely among others from the hotbed of Hymenoptera (ant, wasp, and bee) researchers at Georgia. Our primary goal was to evaluate the pain scale for accuracy and reliability. Stings of many different species have the same numerical value; this does not imply that they are identical in feeling, but that they fall into the same general range of painfulness and presumed effectiveness as predation deterrents. Pain that arises at, or near, the sting site hours or days after the initial sting pain has receded is not considered for this pain scale because it is caused by immunological or physiological reactions to the venom or its damage.

Once it was developed, the pain scale opened possibilities to delve into the secrets of the lives of stinging insects and to predict how their weaponry opened opportunities for them. Predictions operated both ways—we could predict the sting pain based on the appearance, behavior, and life history of a given insect, or we could predict lifestyles on the basis of the sting pain. For example, colorful solitary wasps and bees would be expected to pack a more painful wallop compared with more drab wasps and bees. The reasoning goes that, in evolving a colorful appearance, the option of inconspicuousness, a primary defense masterfully employed by most insects, was largely abandoned. Why should this time-tested defense be abandoned? Perhaps because the insect's life history requires it. The cow killer, *Dasymutilla occidentalis,* illustrates the problem. Cow killers reproduce by locating the nests of other large wasps, entering their nests, and laying eggs in the cells of developing host wasps. The cow killer larva then feeds on the host to complete the life cycle. The problem to overcome is finding enough suitable hosts in the environment in which hosts are rare and usually sporadically distributed. The cow killer must devote much of its daytime activity to searching for these hosts. To make matters worse, female cow killers are wingless and must crawl to locate their hosts. As a result, cow killers are very long lived for an insect, living an entire summer to up to one and a half years. During this long life, cow killers actively crawl around during daytime in plain view of a host of lizards, birds, and other large predators. What chance would a tasty and undefended cockroach, cricket, or caterpillar have surviving for a season or a year under these conditions? Not much chance; those species likely would go extinct. Cow killers are far from extinct: ask any rural Southern inhabitant in the United States. We would predict that cow killers' survival depends on delivering a powerfully painful sting. Indeed, they do, as witnessed by my student who got careless while feeding cow killers and ended up in the university student infirmary.

Cow killer stings are not simply an attention-grabbing 2 on the pain scale, they are a solid, unforgettable 3.

In the Arizona Sonoran Desert, we witness amazing explosions of cactus bees, *Diadasia rinconis.* These drab grayish-brown honey bee–sized bees nest in huge aggregations, often by the tens of thousands, and race around collecting pollen from the abundant yellow, red, or magenta blossoms of various prickly pear and cholla cacti. They are hard to spot and blend effectively with their environment as they suddenly appear in a flower, only to be gone in the blink of an eye. Birds and other predators have difficulty seeing, tracking, and catching these cryptic lightning flashes. Their other effective defense is a short life span. They live only the few weeks during the cactus bloom. They need to evade predators only a short time, not months or a year as does the cow killer. Given this life history, we would predict that their stings are not really needed and would not be especially painful or effective. I know of no one who has ever been stung by a cactus bee, with the exception of entomologists who inadvertently pinch one between an insect net and vial while trying to capture it. Even then, it is hard to get stung. Many entomologists simply reach into the net, grab the bee, and, plop—into a jar it goes. Steve Buchmann, a leading expert on cactus bees who has earned the moniker "Buzzmann," reports that the sting is pretty trivial, a 1 on the pain scale, hardly meriting discussion. My own experiences were similar to Steve's. When I tried to cram scores of netted bees into a wide-mouth jar for venom collection, I received a couple of stings to the side of my index finger. A sharp but mild 1 on the pain scale.

A sting rating 4 on the pain scale is something to be avoided if possible: level-4 pain takes command of one's body and sensory system, shutting down most self-control in the process. Excruciating is an understatement. Fortunately, few insects deliver level-4 stings. This level of sting pain will be described in more detail in the chapters on tarantula hawk wasps and bullet ants.

As more types of stinging insects were investigated, more patterns fell into place. Sting pain, it appeared, really did affect the lives of stinging insects. It affected their lives because of the effect on both actual and potential predators. Predators, parasites, and diseases are the main driving forces in the lives of any animal. According to W. D. "Bill" Hamilton, the great twentieth-century English theorist and naturalist, predators, parasites, and diseases, along with variable environments, are responsible for the evolution of sex.[3] Maybe we should be grateful to our predators and parasites. To an extent that they could think in such terms, stinging insects also should be paradoxically grateful to their predators. Without predators, many opportunities afforded stinging insects would disappear, inherited instead by other adventitious, nonstinging species. Predators opened niches in biological opportunities to those tough enough to exploit them, particularly stinging insects.

Insects, such as cow killers, can live openly and conspicuously because of their painful and effective sting. Stings are not the only solution for insects with conspicuous behaviors, but their stings are effective. Blister beetles, including the Spanish fly beetle, employ another solution. They produce deadly cantharidin, a chemical that blisters skin, mouth, and stomach on contact. Furthermore, blister beetles accelerate the delivery of cantharidin through reflexive bleeding in which beetle blood laced with cantharidin is hemorrhaged through preweakened membranes on the body. Most predators that taste blister beetle blood get the message quickly and reject the beetle unharmed. However the message is delivered—sting, toxic blood, or otherwise, life is always better if tangible message delivery can be avoided. Throughout the animal world, strong, dominant individuals communicate subtle and not-so-subtle messages to weaker individuals not to challenge.

The stronger animal will win the contest, but is it really a win? If the dominant male sea lion loses a pint of blood to defend his harem and the loser loses a gallon of blood, does anybody actually win? Both would have won more—or lost less—if the battle had never occurred. Threat displays evolved to obviate such hopeless battles. Likewise, a cow killer or blister beetle benefits if it is never assailed by the lizard and need not waste its valuable defensive resources or risk injury or death. Nature has evolved truthful advertising slogans for stinging insects to tell predators to "stay away, don't mess with me." These messages can be the eye-popping aposematic color patterns of red and black, orange and black, yellow and black, white and black, or simply any of those colors richly displayed alone. Or the message can scream by way of a rattle, a snap, a squeak, or a rasp. These acoustic warnings are broad ranged, usually low frequency, and generalized so that any predator capable of hearing not only detects the signal but also recognizes that the signal is not a specialized communication among individuals of the same species. These messages are designed not to be confused with a bird's courtship song or a lustful katydid's chirping. Some predators, toads and frogs for example, might not obviously respond to color or sound signals. In these cases, the most basic of all sensory systems—taste—can be targeted for messaging. Nasty tastes in food signify nasty consequences if eaten. Giant velvet mites, squat, red, furry balls on stubby legs, emerge during the first summer rains and wander around looking for winged termites for dinner. Toads, horned lizards, and other known potential predators fastidiously avoid these bulky honey bee–sized mites. They convey their message through their bright red color and, especially, their nasty taste. Lizards will lick a mite and reject it. One sampling seems sufficient for a lifetime. Toads sometimes eat a single mite; even these slow learners get the message and will not eat more.

My curiosity demanded answers to the mystery of why velvet mites were so noisome. From a biological perspective, humans are simply big generalist predators, scavengers, and herbivores that eat almost anything, incorporating a wide variety of animals, plants, and fungi, live or dead, into their menu. Our taste reflects this generality and is tuned to respond to a host of gustatory chemicals that might convey "food" versus "poison." Our taste response approximates the taste responses of other generalist predators, such as birds and lizards, wishful predators of big, juicy velvet mites. If lizards and toads can taste and respond to (reject) velvet mites, should not I likewise detect and respond to their taste? I delicately approached the problem, remembering the childhood admonishment not to eat anything I knew to be unsafe. Velvet mites might be toxic. They might cause blisters. Therefore, they should not be eaten outright. I placed a fat velvet mite on the tip of my tongue, the safest and most distant part from my throat, smashed it against my incisors, and chewed as best one can with incisors. The taste was truly amazing, stunning is perhaps a better word. After the 2-second analysis, I spat out the red juice like chewed tobacco. However, the flavor did not spit out with the crimson fluid. It was bitter, more bitter than quinine or any medicine I had ever tasted. It was also hot and burning, like a habanero pepper. Worse, it attacked the back of my throat and lingered there, that combination of bitter and corrosive heat. I was used to most nasty tastes leaving shortly after being spat out. Not this one. It lingered. The lingering seemed to last forever, an hour at least, before finally releasing its grip.

A stinging insect benefits from communicating its unsuitability as dinner to any predator and in any possible manner. Strutting is a time-tested way of saying, "I am tough. I know you are watching, and you don't want to mess with me." Tarantula hawks and many other spider wasps strut on the ground while frequently flipping their wings. The message is clear, "I want to be seen, and you want to remember how I move so you don't make a mistake and misidentify me."

Visual systems of humans and other vertebrates are geared to recognizing walking gaits of other humans or potential prey or predators. Recognizing walking gaits is an ancient ability of the brain that relies on peripheral vision and does not require clear or focused vision. I frequently search sandy areas of Florida or Arizona deserts where harvester ants are abundant, on the surface searching for seeds to cache. They are about the same size and same color as many velvet ants. Many small velvet ants, diminutive relatives of cow killers, are orangish and about the same size as large ants. Harvester ants typically are a thousand times more abundant on the soil. Yet, out of the corner of my eye, I spot the motion of the one velvet ant mixed with the hundreds of harvester ants. The gait of the velvet ant catches my attention, not its size, color, or slightly different body shape, features my peripheral vision cannot distinguish. They simply move differently than harvester ants, a motion I subconsciously recognize. Harvester ants, powerful stingers in their own right, walk in a jerking fashion. They move a few steps, abruptly stop, move again, halt or slow again, repeating these random movements. This gait might be a warning that the ant is "spicy."

Warnings and gaits all serve the purpose of reducing attacks on defended and stinging insects. Ignoring the special situation of Batesian mimicry, these warning signals are only possible and effective because they advertise real defensive ability. Combined, the artillery of the sting and the sensory trumpeting of warning allowed some stinging insects to exploit otherwise forbidden areas such as desert surfaces, open fields, and even our picnics. Without its sting, the common yellowjacket would be unable to steal ham from our sandwich or imbibe the sweet juice from our peach. Without the sting, the evolution of sociality, especially higher eusociality in ants, wasps, and bees, would not be possible.

Evolution of Sociality

Woe to the animal with poor defenses that casually aggregates with others of its kind. An even greater woe would be to form a society of individuals cooperating to perform community tasks and raise their young together. Predators would quickly devour the delectable, defenseless society, and the story would end with few potentially surviving individuals, no societal reproduction, and no society. Evolution of societies would be prevented, and we would see no social insects or other social animals.

But we do see many social insects and some social vertebrate animals, including humans and naked mole rats. How can this be? All social animals have effective defenses to blunt the attacks of predators. Humans lack claws, horns, or long, sharp canine teeth and cannot run that fast for our size, but we do have big brains and agile hands and arms. Our brains allowed us to tame fire, a defense no other animal uses and that predators of humans fear. Our brains also allowed us to develop tools and weapons. Our hands and arms then allowed us to accurately throw objects or spears at potential predators (something even chimpanzees cannot do), thereby providing defense from a distance. Defense from a distance gave us an unusual and highly effective defense, something lacking in antelope, elephants, and even lions. In essence, we evolved defenses that rendered us nearly invincible and allowed us to become obligatorily social.

Like people, all other social animals have evolved *some* effective defense(s) against predators. Some defenses are structural, as in the case of mole rats that live in tunnels they dig in rock-hard African ground. Predators simply cannot dig them out of their underground fortresses. Termites use similar structural defenses; they inhabit the soil, wood, or hardened self-made nests aboveground or in trees. The effectiveness of these hardened termite nests, found in Australia and in Africa, is apparent to anyone who has ever kicked one.

The other way social insects defend themselves is by active physical defenses that can cause harm. Some aphids, whose soft bodies are pathetically hopeless for defense, evolved sociality through the evolution of a special caste of warrior aphids. These tiny aphid warriors effectively use their sharp beaks to puncture and inject venom into predators that breach the protective plant gall they live within. Most social wasps, ants, and bees evolved sociality largely because they also evolved effective stinging defenses. Granted, some social wasps, bees, and ants lack effective stings, but, in all cases, these species form tiny colonies and live underground or within small hidden nests, or they secondarily evolved the loss of the sting after evolving sociality. The prime examples of secondary loss of the sting are many ants and the stingless bees. The nature of predation pressure against nonstinging ants changed from primarily large predators to other small predators, mainly other ants. Agility, sharp mandibles, and chemical defenses, such as formic acid, proved better defenses than stings against other attacking ants. Stingless bees also possess powerful biting mandibles in association with noisome defensive chemicals and produce wax and resinous structural and chemical defenses against small predators. The sting is not essential in nonstinging ants and bees. Their mandibles, chemicals, and agility provide effective defenses against large predators, as anyone who has messed with a large carpenter ant or stingless bee colony can attest.

The issue is not why some highly social wasps, ants, and bees don't sting but how sociality evolved in these groups in the first place. To evolve sociality, a species must evolve good defenses against predators hell-bent on making a meal of the "wishful society." Why do we not see social grasshoppers, beetles, or flies? Yet we see myriads of social Hymenoptera? The answer lies in the lack within grasshoppers, beetles, and flies of a meaningful defense against large predators. Ants, social wasps, and bees, in contrast, evolved from ancestral sting-bearing wasps preadapted for defense against large predators. As I argued in 2014 in the *Journal of Human Evolution*, a key component for the evolution of sociality in ants, wasps, and bees was the evolution of venom, and some modification of the stinger and behavior.[4] This sting-venom evolution then allowed social evolution to progress despite powerful predators working against that social evolution.

SWEAT BEES AND FIRE ANTS

There may be many surprises in store for us when the life-histories of these seemingly monotonous and uninteresting bees [sweat bees] have been subjected to more careful scrutiny. —William Morton Wheeler, *The Social Insects,* 1928

The ferocious little pests. —Edward O. Wilson, foreword, *The Fire Ants,* 2006 (in reference to the fire ant *Solenopsis invicta*)

Sweat bees, those rascally denizens that appear during the peak of hot, humid, sticky summers in eastern and central North America, are familiar visitors to backyards and social gatherings. Why "sweat bees," a peculiar name, for sure? Do they sweat? No. Do they make us sweat in fear? No. Are they even bees? Yes. OK, that part of the name makes sense, but where did "sweat" come from? It originated from the unusual habit of some species of these bees: they land on and lap sweat from human skin. Most kinds of bees do not collect sweat, which makes the sweat bee's habit oddly noticeable.

Bees are a collective of 20,000 species, outnumbering all warm-blooded animals on Earth.[1] Sweat bees are members of the enormous bee family Halictidae, which contains 4,387 species, more than all mammals, excluding bats. Sweat bees live throughout the world on all continents except Antarctica and display a greater diversity of social behaviors and life histories than any other group of insects. Many are strictly solitary, that is, lone female bees work in isolation, collecting food, building a nest, and otherwise providing for the young. Others remain solitary, but nest in aggregations of many individuals, each working on its own within the close aggregation of nests. Some species are semisocial, with two or more females working together in the same nest but performing different tasks. Finally, some are truly social, with several individuals living in the same nest, including an egg-laying queen and workers. To complicate matters further, some sweat bee species are solitary at one time or in one location and social at other times or locations.

Most sweat bees are small, 3–12 millimeters (⅛–½ inch), and black, grayish, or metallic green or blue, sometimes with splashes of yellow or red. They usually nest in the ground. Nests consist of a tunnel excavated from the soil surface, descending into the earth, usually with side branches leading to individual cells where the young are reared (only females do the work; males sit around and, at most, guard nest entrances). Once a cell is prepared, the female collects pollen and nectar from flowers to form into pollen "balls," or "loaves," the sole food for the young. She then lays a single egg in each cell on its pollen provisions and seals the cell before working on the next cell. Curiously, female sweat bees, as are most other bees, wasps, and ants in the order Hymenoptera, are haplodiploid organisms, in which fertilized eggs become female young and unfertilized eggs become males. This allows females to choose the sex of each young, something people cannot naturally do (perhaps a good thing overall). In practice, this female choice often translates into males getting the short end of the loaf in the form of smaller food provisions; hence, males are often smaller and scrawnier than females.

Sweat bee life begins as an egg laid on or near its doughy pollen mass. In a few days, the egg hatches into a tiny, almost transparent whitish larva that feeds on the pollen. In the process, it grows larger, molting apparently four times into a successively larger grub-like larva. Sweat bee larvae lack a connection between their midgut (stomach) and their hindgut, rendering them incapable of defecating, which is likely a good thing because each is confined to a small cell, feeding on a rich, potentially spoilable food source. At the end of larval feeding, the connection forms between the mid- and hindguts, allowing the now enormous larva to make its only, and probably much-appreciated, defecation. Unlike some bees, the larva does not spin a silken cocoon; instead, it molts within its cozy, wax-lined cell into a pupa, the resting stage in which the adult bee develops. Pupae are delicate creatures, and usually this stage is short. The pupa molts into an adult that remains in the cell during unfavorable seasons until the winter or adverse periods are over, then digs out to become a free-flying adult. Adults often live a comparatively long time for small insects, giving them opportunities to visit many flowers, typically a succession of different floral types. Some sweat bees have two or more generations per year, others simply one generation. In any case, males and females mate, and females are able to store sperm for later use after the season with males has passed.

The insides of sweat bee cells are lined with a waxy protective coating secreted by the mother bee.[2] This impermeable coating within the cell protects against problems, including excess wetness from rain, desiccation during dry periods, and fungal or other pathogen problems. The lining is produced in the Dufour's gland, a curious gland associated with the sting, and is applied to the insides of cells with a combination of the abdominal tip and the tongue. The gland name, itself, also has a curious history. It is named after Léon Dufour, a prominent French physician, scientist, and scholar. Dufour mentioned in 1835 that a "plastique" cell lining in some bees appeared to be derived from a large abdominal gland.[3] In 1841, he elaborated that the glandular secretion also was used by females to coat eggs. Along the way, the gland acquired the strange name "alkaline," or "basic" gland. It was believed that its fluid was alkaline in nature. Even though this characterization was disproved, the term "alkaline gland" persists. When the gland became known as Dufour's gland remains a mystery. Dufour never named it, and after a few brief forays into the subject moved on to other topics. With numerous synonyms (e.g., sebifique gland) in use, somehow the name Dufour's gland arose, perhaps shortly after 1841, and stuck, apparently in honor of the great man. Odd, how today he might be best known for this obscure gland named after him.

Even though more than 4,000 species of sweat bees exist and that some species lap human and animal sweat, we still don't know why they collect sweat. Surprisingly, little research has been conducted to address this question, and most of what is known was performed in 1974 by Edward Barrows, at the time a graduate student at the University of Kansas. Ed showed in a series of choice tests that sweat bees are attracted to and prefer table salt solutions over controls that lacked salt.[4] Because salt does not evaporate or have any odor, something else must attract the bees. Likely candidates might include lactic acid, carbon dioxide, or the mosquito attractant 1-octen-3-ol, all released from skin surfaces. We simply don't know. We also are left wondering whether bees are seeking salt, water, or some other sweat component as a nutritional requirement; lactic acid, octenol, or carbon dioxide seem unlikely. An additional complication is that not just sweat bees collect sweat. Africanized honey bees on occasion collect sweat. In Asia, stingless bees in the genus *Trigona* collect sweat and are sometimes called sweat bees. In parts of Africa, stingless bees that collect sweat are often called sweat bees and are sometimes called "mopane flies," even within academic circles.[5] Yet none of these insects are either true sweat bees or flies. These African stingless bees *are* actually stingless; that is, they lack a functional sting. Not that stingless bees are defenseless. They have sharp mandibles and attack in mobs, biting eyelids, nose, and ears and crawl into ears, nose, and mouth, a most definitely unpleasant experience.

Although some Australians call them "sweet bees," a much more pleasant moniker than sweat bees, we appear to be stuck with the term "sweat bee" for these little bees that visit our summer activities. These visitors, unlike stingless bees, do not bite, but they do sting. In a typical scenario, a person is leisurely enjoying a pleasant July afternoon, relaxing outdoors with a favorite drink in hand. A few flies are buzzing around, and the occasional honey bee visits a nearby flower; otherwise, the kids are playing, and the afternoon is perfect. Up to one's mouth comes the drink and, *ouch*, something stung me! Tranquility is ruined by a little dark bee in the crook of the elbow. The little bee meant no harm; it was just enjoying one of its favorite beverages, the sweat accumulated in the fold at the elbow between the forearm and upper arm. When the glass was raised, the bee became pinched between the skin. The threatened bee responds defensively by stinging in an attempt to stop the pinch. Often that strategy works. Other times the unsympathetic recipient of the sting responds by smashing the poor bee.

The pain caused by the sting is nothing serious. It's pure, clean, and tidy, like "a tiny spark has singed a single hair on your arm." The pain is not likely to get you serious sympathy; that is, unless one is a small child, in which case no excuse is really needed for hugs and sympathy. The pain goes away momentarily and leaves little evidence of its occurrence. The sting rates a classic 1 on the sting scale, thereby providing a convenient reference for comparing other past and future stings. It certainly does not in any fashion compare with the pain of a honey bee sting, rated 2 on the sting scale.

Fire ant. Now there's a nasty animal. Unlike sweat bees that mean no malice as they sweetly go about their business pollinating flowers we love and foods we enjoy, fire ants have an attitude. Humans also have an attitude toward fire ants as expertly expressed by Walter Tschinkel: "Most people hate fire ants without reservation, without reflection. Perhaps that is what the fire ant has to offer us—something we can all agree to hate, something about whose reprehensibility no argument can be made."[1] For their part, fire ants' first response to contact with human flesh is to sting. Fire ants have not learned how to make friends and influence people. Who are these fire ants? What do they do? Where did they come from? Why are they so nasty? How do I get rid of them?

Fire ants are small polymorphic (many different-sized individuals in one colony) ants belonging to the largest and most successful subfamily of ants, the Myrmicinae. They are all in the genus *Solenopsis*, a 185 species–strong group that has frustrated some of the greatest ant scientists, including Carlo Emery, William Creighton, William Buren, and Roy Snelling. The frustration arises in large part because the smaller worker ants, called minors, are amazingly similar among species to most anyone who has not devoted much of their life to ant taxonomy—and even to some who have. The comment regarding ants in general made by William Morton Wheeler in 1910 that "the species of ants often differ from one another by characteristic too subtle and intangible to be readily put into words"[2] applies especially to fire ants. For easiest identification, major workers, the largest "big headed" individuals, are required, and they constitute a minority of the overall population. Imagine the fun of searching for the largest ants while picking through a swarming mass of outwardly oozing stinging ants and the problem becomes clear. The other problem in assigning names to fire ants is that over the years too many experts have further confused the already-muddled situation by making unnecessary changes, causing an exasperated Roy Snelling to comment about one author: "The best method of dealing with this volume is to disregard entirely the statements of the author."[3]

Fire ants are loosely grouped into the fire ants proper and the "thief ants," the latter category with more species by far. Thief ants make their living by stealing from other ants. These tiny ants build their nests in conjunction with other ants and proceed to dig tiny tunnels that join into the chambers of the other ants. From these tunnels, worker thief ants raid and steal brood (immature ants: eggs, larvae, and pupae), which they take back to their own colony to consume. They succeed partly because their tunnels are too small for passage by the host ants. Thief ants cannot sting humans and are invisible to all but a tiny portion of people who actually care about them.

Ants in the other category, true fire ants, are larger and of more concern to humanity than thief ants. These ants all sting, all are nasty, and all originated in warmer areas of North America and South America. Their behaviors, like their general appearances, are similar. If you've met one fire ant, you've met them all. All fire ants build large colonies of thousands to hundreds of thousands of ants. They are all polymorphic, with a range of small to large ants, and eat almost anything that comes their way and provides calories—live prey, dead animals, seeds, nectar, honeydew, or other plant material. Fire ants aggressively defend their territories and attack intruders. They all sting, and those stings hurt. Fire ants probably all form into balls that float on the water's surface when floods come. Six species can be found in North America. Three, the southern fire ant (*Solenopsis xyloni*), the pleasantly named golden fire ant (*S. aurea*), and *S. amblychila* are long-established natives. Another species, the tropical fire ant (*S. geminata*), might be native to the United States or might have migrated naturally or with assistance from humans many centuries ago. The final two fire ant species are the "imported" fire ants (a peculiar name that implies they were introduced by desire), *S. invicta* and *S. richteri*. These two were both transported from South America to Alabama in the first half of the twentieth century.

My first encounters with fire ants came in the 1970s when I was a graduate student at the University of Georgia. I had vaguely heard horror stories in the news of these invading ants from South America who were abusing southern hospitality. Otherwise, I knew little about them. What an opportunity to see them up close. My first reaction was disbelief. These ants are tiny, not at all like the big, self-respecting ants I was familiar with. From their reputation, I expected something enormous. The second impression came quickly as I brushed some loose soil from an ant mound to look inside. The ant reaction was swift and decisive. They immediately crawled onto my hands and up my arms, many stinging, as the horde moved ever farther upward. These ants are nasty. They don't just bite, a minor pinch delivered by most of the ants I had known, they bite and *sting*. The name fire ant accurately describes their stings. When dozens sting at once, a habit for which they are famous, that unlucky part of the anatomy feels on fire. The bite is inconsequential.

I am not sure I have ever felt the bite. The sting overwhelms the bite sensation, which is lost in the surrounding sea of pain.

Although fire ant stings are unpleasant, fire ants are fascinating. The fire ant life cycle starts out as typical for most ants, with winged virgin males and females (queens). They embark on mating flights during warm, pleasant spring and summer days in which thousands engage in the mating ritual, actually a mating frenzy, with airborne pairs grappling, falling to the ground, and mating. This is furious sex in the fast lane, with only one chance to get it right. Males and females mate for 10 seconds, once in their lifetime. Males need to act quickly. They die or are killed and dragged off by other ants within hours of leaving the nest. The mated females break off their wings, often leaving a chaff of discarded wings blowing in the breeze, and scurry about looking for a suitable nest site. Time is precious. Ant queens are universal desserts for most small or large predators. Once a nest site is found, the queen, sometimes joined by other queens, digs a short tunnel in the soil, seals the tunnel entrance with soil, and makes a cell in the bottom. Here she raises a few tiny "minim" workers entirely from her body reserves of stored fat and metabolically cannibalized wing muscles. Once they reach adulthood, the minims go on raiding parties to steal brood from other queens. Along the way, many queens die or are executed, resulting in a few surviving micro colonies. The minims forage for food, rear the first normal-sized workers, feed the queen, and take over all nonreproductive colony activities. The colony is on its way.

Returning to the nitty-gritty of fire ant sex, each mated queen receives only about 7 million sperm with her one mating. She has to raise millions of workers from one 10-second mating and 7 million sperm. She must use her sperm wisely and sparingly if her colony is to thrive, live long, and reproduce. She uses only about 3.2 sperm for each successfully reared worker or virgin queen[1] (and some larvae are eaten or otherwise die before adulthood). Compare this to human use of about 100 million sperm per offspring, assuming fertilization is successful, and the efficiency of the fire ant queen becomes obvious. The queen's small colony grows from a handful of workers to a thousand or so in a year, to tens of thousands by year 2, to nearly 100,000 by year 3, and to 150,000 by year 4. Final colony maturity occurs around year 5 and stabilizes thereafter at around 200,000 to 300,000. The colony dies at a ripe old age of 5.5 to 8 years because the queen runs out of sperm and cannot produce more workers.[1] For a queen, it pays to be frugal!

An individual ant, whether it's a worker, a queen, or a male, starts as a tiny egg laid by the queen. This egg, like the rest of the 2 million to 3 million eggs the queen will lay over her lifetime, hatches into a tiny larva, a translucent, whitish blob that lacks legs and the ability to defecate. However, it can, and does, readily eat the food provided by attending workers, growing over a thousand-fold in weight as a result. Periodically, during this growth, the larva molts, shedding its undersized skin in exchange for a larger skin in which to continue growing. Finally, the larva arrives at its terminal larval instar and eats its last meal. Meanwhile, it has not defecated during this whole process. Like the sweat bee it has no connection between the hindgut and rest of its digestive system. Perhaps this connection did not form earlier to prevent contamination of the nest. When the connection finally develops, the larva takes an enormous dump, called the meconium. (We can only imagine how this feels to the larva.) What happens to the meconium? Specialist workers rapidly collect it, lick the oils from it, and then deliver the oils to the queen. This adds new currency to the old saying, "What goes around, comes around." The dried meconium is discarded. Apparently, within its oils are the precursors to juvenile hormone, a stimulant that activates egg production in the queen. The now shrunken larva molts into a pupa, the resting stage during which the adult ant is developing. The adult ant emerges from the pupa and takes up its role as part of the colony "superorganism." Workers live several months before dying, often from the challenges and risks of foraging; meanwhile, the reproductives wait for their magical moment—the mating flight.

Like sweat bees, fire ants also have a mysterious Dufour's gland. In sweat bees, the Dufour's gland serves as the painter and sealer of their larval home. In fire ants, the Dufour's gland is the chemical jack-of-all-trades and the master of chemical communication. When a fire ant forager discovers a bounty of food, she runs back to the colony dragging her abdomen. Shortly thereafter a group of recruits leaves the nest, following the trail left by the abdomen-dragging ant. The trail they follow originated from the Dufour's gland of the successful forager. She only needs to lay a trail of 0.1 picogram per centimeter for the nestmates to follow. That is less than 1 divided by 3 followed by 14 zeros of an ounce. In addition to making a trail to follow, the Dufour's gland activates ants to follow, something analogous to getting the donkey's attention before it will move.

Man's best friend may be the dog, clearly not the fire ant. In contrast, the fire ant's best friend is man. In the words of Walter Tschinkel, the grand master of knowledge about fire ants, "If fire ants have a religion, humans must surely occupy the position of God, preparing a place for them."[1] To appreciate why we might be gods to these little creatures, let's examine the ant's life. Fire ants like soft soil, especially sandy soil that is easy to dig. They also like warm, sunny areas, preferably those with grass intermixed with other plants. The ideal fire ant habitat contains a variety of food resources, insects, and other small prey, along with seeds and other plant sources. The ideal habitat also has few other ants competing for space and resources. Such areas are disturbed habitats, or what ecologists call successional habitats, in which no established species strongly dominates for long. These areas are perfect for tramp species and other weedy species, including fire ants. In essence, fire ants are simply six-legged weeds. In nature, disturbed areas are relatively scarce, occurring mainly after fires, floods, powerful storms, major pest/pathogen outbreaks, or huge tree falls that dramatically alter an area. Humans raze land to plant crops, to graze cattle and other livestock, to culture grass and lawns around their dwellings and spaces, and intentionally or unintentionally, to burn areas. De facto human activity maintains these areas in a state of permanent disturbance. Human disturbances reduce or eliminate many competing native ant species, making these areas perfect for fire ants.

Fire ants share properties with many other weedy species. They grow rapidly, reproduce prolifically, quickly invade disturbed areas, and fiercely compete with other species. Humans supplement our regular activities that produce the disturbed areas ideal for fire ants by giving fire ants another special gift—insecticides. We declared war on fire ants with our insecticides. The great naturalist E. O. Wilson once described the insecticidal war on fire ants as "the Vietnam of entomology."[1]

The war started on a small scale in the 1940s with government agencies dumping calcium cyanide on fire ant mounds in attempts to eliminate them. The attempts failed. Next, with no scientific background to support the operation, the wonder insecticide chlordane, a nasty, persistent, environmentally destructive chlorinated hydrocarbon, was added to the arsenal. With no meaningful success (the ants were spreading like wildfire), the chlordane club was replaced with the newer clubs of heptachlor and dieldrin, also nasty chlorinated hydrocarbons. Surely, these should solve the problem. Again, as should have been expected based on a lack of good science or understanding, the fire ant prevailed and continued relentlessly expanding its range and population. The environmental effects were so disastrous, except for the fire ant, that the program was modified and then halted. Ah, yet a better club was needed. Enter Mirex. As the reader can now expect, this program, too, failed miserably, and the ant continued thriving and enjoying the fruits of human labor. By the mid-1970s, the Mirex adventure was in its last gasps. About this time, the target of this war, the "red" imported fire ant was described as a new species by William F. Buren, a mild-mannered, soft-spoken gentleman generally given to understatements. Bill named the species *S. invicta*, in reference to the Latin "invincible," apparently his way of responding to the failure of the fire ant wars.

Have we won the war with the fire ant? Not by a long shot. The score is in, and the fire ant is winning. Extensive air bombing of large areas of the South with poisonous insecticides failed to eliminate the ants. In fact, our actions helped fire ants to reduce or eliminate competing ants that, after decimation, are not able to recolonize as quickly as fire ants. Mowing lawns also failed to rid lawns of ugly mounds or the unpleasant ants; they only succeeded in splattering soil around and shortening and broadening ant mounds. We have also failed to control the spread of the fire ants to new parts of the country. Fire ants are now present in Southern California. We have had a few successful skirmishes in this war. Imported fire ants got footholds in Yuma and Phoenix, Arizona. Both introductions were eradicated, perhaps with the aid of a hot and dry climate, conditions not conducive to fire ants.

The failure of humanity's war on the fire ant left homeowners in the South without weapons against the ants in their yards. About this time, Walter Tschinkel at Florida State University in Tallahassee formed a research group called the Fire Ant Research Team, complete with a logo featuring a menacing-looking fire ant standing over the Florida state capitol circled by the motto "Today Florida, Tomorrow Dixie." One of the tasks of this distinguished team was to provide some relief to the beleaguered homeowner. In a viscerally satisfying way, the team did score a small victory over individual fire ant colonies. The solution was a simple, safe, nontoxic, and highly satisfying means of killing colonies, and it cost nothing, spare a penny perhaps for energy. The technique is simple: boil 3 gallons of water, select your favorite fire ant colony, and slowly pour the 3 gallons of water directly down the center of the mound. When done carefully with minimal run off, the hot water penetrates deeply into the soft mound, killing not only most of the adults and brood but also the queen. With a high success rate, this method provides temporary relief and the inherent satisfaction that a small environmentally benign victory was achieved.

The point of the digression into the fire ant wars is to illustrate how people are the fire ant's best friend. Our everyday activities greatly helped the fire ant, and our insecticidal overexuberance ensured its success in its new home. How could this be? The fire ant, like any other colonizer or invader, must battle with any current residents for space, food, and other resources. If the resident population is large, strong, and established, the battle spoils usually fall to the residents. If the residents, in this case other ants, are eliminated or weakened, the invader's job of colonizing becomes much easier. The insecticidal wars did just that. They eliminated most of the established resident ants (along with any fire ants that might have been struggling to make it). The playing field was leveled for fire ants or, should we say, prepared in their favor. With little or no competition from native ants, the reproductive fire ant queens that settled in the vacated land had excellent chances for success. This scramble is lopsided in the fire ant's favor. The originally established ants in the habitat generally produce fewer reproductives, and their reproductive flight season is usually short. In contrast, invading fire ants, as is classical for tramp species, produce prodigious numbers of reproductives, and their reproductive flight season is long, typically covering the entire warm part of the year. These traits give the fire ants a powerful heads up over native ants in the struggle for the newly cleared areas left after the insecticidal depopulation is over. Soon the treated areas are teeming with thriving fire ant colonies. With humans as their best friends, fire ants need not worry about their enemies.

In fairness to the fire ant, we should say that it can be man's friend, albeit perhaps not the best friend. Humans like disturbed areas, such as fields and pastures, where crops and domestic animals are grown. Within the rows of crops or pasture spaces live crop and animal pests that compete with us for the spoils of our labors. Fire ants that inhabit these fields and pastures can be useful, maybe even welcome, predators of pests, such as the sugarcane borer, a caterpillar that destroys sugarcane plants in Louisiana; boll weevils and pink bollworms that ruin Texas cotton; mosquito eggs in floodwaters of rice farms; and horn flies and stable flies within cow patties, a serious problem for cattle. These are just a few examples of how fire ants help us. Perhaps these are not sufficient for us to raise our champagne glasses in toast to the fire ant, but they do provide a positive side to one of our newest acquaintances.

I am often asked how one can distinguish fire ants from other ants. In the South, the heartland of imported fire ants, a microscope or fancy identification guides and taxonomic keys are not needed. All one needs is a good tennis shoe; that is, a tennis shoe on a good foot. I call this the

"Nike" test. Simply walk up to the mound in question, give a good swift kick with the heel of the shoe to knock off the top soil of the mound, and quickly step back a few steps. If the top of mound turns black within 10 seconds with roiling ants, these ants are fire ants. At this point, take a few more steps back. The test is not without its risks. If not efficiently executed, a few ants might remain on the shoe, and these ants will inevitably crawl up the ankle looking for somewhere to sting. Stomping and fast brushing will usually solve this problem.

People west of the Pecos River in Texas through New Mexico, Arizona, and California often think they have no fire ants. To dispel this myth, try the backyard barbecue test. It's simple. After a warm, midsummer afternoon barbecue, when you've finished enjoying the chicken, toss a bone or two into the yard. To be really fancy, cover the bones with a rock or a scrap of wood. Then, early the next morning, as the sun is glowing in the east, inspect the bones. They will likely be teeming with little light-colored ants. In many cases, these little ants are native fire ants. They are similar to the imported fire ants, only less numerous and less obvious because the hot, dry climates in western areas keep the ants underground during the day where they are not noticeable. They are less numerous because, as native ants, their populations are more in balance with other species of ants, unlike the runaway population explosion in the South, where their imported brethren have taken over and displaced most of the native ants. Although less common than the imported fire ants, native fire ants in the West are not meek, mild characters. They are every bit as feisty and ready to sting as imported fire ants, as anyone who has used fingers to remove the morning chicken bone from the yard can testify.

Are fire ants dangerous? Yes and no. No, in that most of us suffer no ill effects of fire ant stings, save the damage to our egos and the rude disruption of our peace and tranquility. The next day, other than some white, pimple-like pustules caused by imported fire ant stings, we are none the worse. A classic illustration of the usual lack of long-term damage from fire ant stings is the saga of an inebriated fellow in Houston, Texas. Here is his experience as described by the treating doctors:

> A 49-year-old alcoholic was brought to the hospital at 2 a.m. on a Sunday morning. After drinking all day and all night on Saturday, he attempted to go to a friend's house to sleep. Arriving at the ditch in front of his friend's home, he was overcome by drowsiness, and in the dark selected a fire ant mound as his pillow. . . . Approximately 5,000 of these [sting] lesions were scattered over his face, trunk and extremities. His vital signs were normal, as were the results of the remainder of the physical examination except for the strong odor of alcohol on his breath. The following morning, the patient had his "usual hangover" but otherwise felt fine.[4]

In fairness, there is also a yes answer to the question, Are fire ants dangerous? The yes refers to unlucky people whose encounters with fire ant stings sadly result in allergy, not happy endings. The tiny percentage of people who are allergic to stings suffer anything from systemic skin reactions to difficulty breathing, a drop in blood pressure that causes fainting or loss of consciousness, and trips to hospitals. Curiously, the rate of allergy to fire ant stings is actually much lower than allergy to honey bee or wasp stings. Allergy to bee and wasp stings is around 1 percent to 2 percent of the population. In contrast, less than 1 percent of the population is allergic to fire ant stings, especially surprising given that about half the resident population is stung yearly in fire ant–infested areas,[5] and only 10 percent or fewer of residents in bee and wasp areas are stung yearly. The reason for the lower rate of allergy to fire ants is unclear but likely related to the much smaller amount of venom protein injected with fire ant stings than with bee or wasp stings. Nevertheless, a handful of people actually die each year from allergic reactions to fire ant stings; fortunately, this is not the norm, as most of us get by with a few expletives and not much more.

The story of fire ant venom and its chemistry rivals any good mystery thriller, complete with murder, intrigue, and detective work. Fire ant venom consists largely of piperidine alkaloids, compounds related to coniine, the main toxic component of the deadly poison hemlock.

Socrates was forced to drink hemlock in 399 BCE following his conviction for impiety, essentially the excuse levied against him for being a rabble-rouser who challenged the ideas of the powerful in Greek Athenian society. Coniine is a water-soluble alkaloid; hence, it is easily brewed into a bad-tasting tea. In contrast, fire ant piperidines are insoluble in water and lack any taste, rendering them poor candidates for making a tea for murder. Their lack of water solubility also means we cannot determine how poisonous fire ant piperidines are to humans because they do not flow through the blood or lymphatic streams from sting site to the heart, lungs, or other vital organs. Instead, these alkaloids stay in the skin, poisoning the local area and causing the delayed pustule formation so characteristic of imported fire ant stings. Equally mysterious is why only imported fire ant venom causes skin pustules. Native fire ant stings don't cause pustules; thereby, they provide a convenient, if not pleasant, way to differentiate between native and imported fire ants in the United States. Could the difference in pustule formation be related to the fact that the alkaloids of native fire ants are smaller than those of the imported fire ants and that these smaller alkaloids are more water soluble and can be carried away from the sting site? Or could native fire ant alkaloids simply be less locally toxic?

Good sleuthing was necessary to understand the chemical nature of fire ant venom. For years, the odd nature of fire ant venom was known. Unlike most ant, wasp, or bee venoms, which are a blend of water-soluble proteins and peptides (small proteins), fire ant venoms form venom droplets that float in water and lack meaningful amounts of proteins. The venom chemistry proved elusive, first with misidentifications in the mid-1960s, followed by a high-profile report that the identifications were, in fact, wrong.[6] The story was back at its start with venom being "an amine."

In the early 1970s, Murray Blum took up the challenge and assembled an able team to tackle the fire ant venom problem. Murray, well known for his collection of tobacco pipes that he could rarely keep lit and for his mean game of squash, was located in the fire ant heartland of Georgia, where he was suited for the challenge. In an extensive series of detailed chemical papers, he and the group discovered that "solenamine," as the active components were called, consisted of an assortment of 2-methyl-6-alkylpiperidines with the alkyl groups ranging from 11 to 15 carbons. The coniine Socrates was forced to drink differs only in having a propyl (3-carbon) group at the piperidine 2 position in place of the 1-carbon methyl for fire ant venom, lacking the 6 position side chain entirely. The fire ant story gets even stranger. The more primitive native fire ants have mainly piperidines with 11-carbon side chains, whereas the two imported species have mainly 13- and 15-carbon side chains.[7,8] Given that the main venom difference between the native and imported fire ants boils down to 11- versus 13- or 15-carbon side chain length, we might conclude that the skin pustules are probably caused by the longer chain lengths in the imported species than in the natives. Oddly, in other studies, it turns out that the 11-carbon piperidines are more toxic to fungi or a variety of bacteria than the longer side-chained components. Given that the venom appears to serve an important role in fire ant nest hygiene by controlling fungal and bacterial pathogens, why, then, would the imported fire ants evolve less effective, and more metabolically costly to produce, venom components? Could it be that increased defensive value against big predators, like humans, is the reason? We don't know. We have many more rich mysteries about the fire ant to solve, as the fire ant does not yield its answers easily.

This brings us to the sting of fire ants. Most of us in the United States assume the worst is here, and often are dreadfully afraid of the stinging fire ants we already have. Never fear, others even worse might be lurking in South America, awaiting us to transport them here. According to reports from distinguished myrmecologists (people who study ants), possibly more painful candidates might include *S. virulens* and *S. interrupta*. Stay posted. But really, the stings of fire ants are not that bad. On the pain scale, a fire ant sting musters a pain level of only 1, pale in comparison to even the common honey bee. The fire ant sting is sharp and immediately sends a burning sensation to the area. The pain, however, lasts only a couple of minutes before receding to the state of "oh, yeah, I can feel it, but it's nothing to get excited about."

7

YELLOWJACKETS AND WASPS

[Yellowjackets and baldfaced hornets] terrorize housewives, ruin picnics, and build large aerial nests that challenge fleet-footed stone-throwing boys the world over. —Howard E. Evans and Mary Jane West-Eberhard, *The Wasps*, 1970

THE YELLOWJACKET. Mental images explode of something bright and flashy, something brash and perhaps rash, something to be watched and considered. Indeed, these images apply perfectly to yellowjacket wasps. They are bright, flashy, and require watching because they can sting. Delivering a painful sting to the oblivious or to someone foolish enough to grab one or to meddle with its nest is a yellowjacket's special talent. Their bright yellow and black coats, indicate that yellowjackets are true masters in communicating their defensive potential or, if needed, their actual abilities to any person or other large visually attuned creature. If the potential assailant has dim vision, yellowjackets apply another warning trick. They buzz loudly with a fierce, readily audible high pitch to a wide range of audiences. This shrill buzz is different from the normal buzz of flight of a yellowjacket, bee, or fly. It is as distinctive, general, and easily recognizable as the rattling sound of an alarmed rattlesnake. The role of the yellowjacket's buzzing and the snake's rattle is the same: to warn the listener to stay away or suffer severe consequences. When captured, common flies attempt to mimic the same high-pitched warning buzz of a wasp or bee, only they're bluffing. Imitating a dangerous animal by converging the sound, color, odor, or behavior of another animal is called mimicry. If the mimicry is fake, that is, the animal poses no actual threat, as in the case of the fly, it is called Batesian mimicry, named for the famous nineteenth-century naturalist Henry Bates, who first described the phenomenon. If the mimicry is honest and the animal actually poses a threat, as among the several different species of yellowjackets that look and sound similar, it is called Müllerian mimicry, after the German naturalist Fritz Müller, who first described in detail this form of mimicry.[1]

When given the chance, all young children are naturalists. Like other children, I was a young naturalist, who, in my case, happened to be enraptured by yellowjackets, and that biggest of yellowjackets, the baldfaced hornet. Their active, seemingly carefree style of flitting about in the warm sunlight was appealing. Their bold colors added an air of excitement. They challenged me to find where they were going and where they lived. Once I located their nest more opportunities were revealed. How close could I get without them noticing? Could I count how many came in and out in a minute? What will they do when presented with a challenge? Once I watched with youthful fascination as my father, a venerated Wisconsin forester brimming with practical natural history wisdom, decided to see whether he could eliminate a yellowjacket colony living beneath the masonry stone steps leading to our porch. He didn't want to kill them by pouring gasoline into their entrance at night, the unapproved, though traditional, way of dealing with yellowjacket colonies, for that would leave a smelly and potentially unsightly mess. Lighting the gasoline, the exciting flaming climax of traditional yellowjacket eradication, was too hazardous and illegal for his style and personality. He also didn't want to poison our front yard with typically foul-smelling, and likely unsuccessful, chemical sprays. The chosen solution was to fill the entrance with mortar to trap the wasps inside. We went out with a red cellophane-covered flashlight (yellowjackets cannot see red) and filled the entrance hole. The next morning yellowjacket colony life was as normal. They had simply dug through the moist mortar at night and continued life as usual. The next night we stuffed some steel wool down the entrance to block it and filled in mortar again. The yellowjackets buzzed softly from within during the process, but they could not attack us. Fooled again. This time the yellowjackets dug a side tunnel through the earth beneath the walkway and exited a short distance away. I learned from this experience that these insects, like people, were good at solving problems and adapting to challenges that nature might bring to them.

My father ensured we did not get stung during these adventures. He knew the painful sting of yellowjackets, and, as a good parent, he wanted to protect me from getting stung. I viewed the operation as a good, clean learning experience but certainly not high adventure. For this, I would tag along with neighborhood boys as we played in the streams, fields, and woods looking for snakes, frogs, toads, worms, or anything else that caught our attention. We were game hunters in the tradition of our long-ago ancestors in Africa, only we hunted small game, and we didn't eat it like our ancestors would have. On a pleasant summer day, one of the older boys came upon the perfect small game, a baldfaced hornet nest. Soon we were throwing rocks in the direction of the nest. And, yes, I was stung. Screaming and yelling were all part of the adventure—as we were all screaming—but no crying was allowed. I licked my wounds in silence.

The yellowjacket episodes instilled several lessons. First, nature is a two-way street. If you act toward another creature, it will respond to your action. Second, we as humans cannot always dominate and predict the consequences of our interactions in the natural world. Finally, nature, insects, and stinging insects, in particular, are thrilling. Children are simply untrained scientists, and scientists are simply trained adult children. Around the age of 12 or 13, we move into the ranks of young adults and are expected to leave childhood behind, except in memories. In junior high school, we learn rigorous academics and trades. My interests shifted first toward math, especially geometry, then to physics, and finally to chemistry. Insects and biology were left behind, but not forgotten, by the boy within me.

Years later my chemistry and entomology training took me to Costa Rica where I studied the ecology, genetics, and defensive behavior of killer bees. In genetic terms, killer bees are simply biological wild-type honey bees that had not yet been inbred, genetically modified, or domesticated. The brilliant, talented geneticist Warwick Kerr had brought them over from the Pretoria region of South Africa by request of the Brazilian government. They escaped captivity, were better adapted to the warm climate of Brazil than the domestic bees previously introduced, and in their northward range extension had reached Costa Rica. They still retained their full range of natural defensiveness against human and other mammalian predators.

On a break from bee work, I and Hayward Spangler, an expert in insect sounds and acoustics, decided to visit Frank Parker, who, at the time, was investigating the ecology of screwworm flies. Frank is a tall, impressive, distinctive man who was locally referred to as "malo Gringo grande" for his friendly manner combined with an unstoppable energy as a human vacuum cleaner sucking up any insects and spiders he could find in the field. His sense of humor extended to taking visitors to the field to help capture screwworm flies attracted to a festering three-day-old mash of rotting pig liver. He got great delight in watching the expressions on the faces of hardened entomologists when he slammed his white insect net onto the pile of rotting liver to capture a fly.

The day we visited Frank, he was in a forest meadow at mid-level elevation along the western slope of a mountain range in Guanacaste Province, Costa Rica. Hayward and I left Frank to his flies and went exploring for any interesting ants or wasps that might sting or make sound. Success was close at hand, about a hundred meters up from Frank's camp. In a small, thorny, nearly impossibly dense bush was a large nest of *Polybia simillima*, a tropical epiponine wasp, independently noted by the early twentieth-century naturalists Philip Rau and, later, O. W. Richards for its especially painful stings and its ability to leave its stinger in one's flesh. Opportunity knocks and then flees. I was not about to let this opportunity flee. Frank and Hayward declined, preferring to stay with the flies, while I donned my bee suit, complete with protective veil, to bag the wasp nest for dissection, venom collection, and analysis. In all my tropical experiences, I had never encountered this uncommon species before and was not about to miss the opportunity. I had learned previously that any black stinging insect is telling me something and needs to be approached with utmost care. These menacing wasps are black, buzz fiercely, attack with speed and agility, and leave their stinger in your skin, all properties I had anticipated. With clippers to remove the spiny branches and a bag in hand, I expected the operation to be routine, and I would have my wasps. Wrong. The lessons from childhood experiences with yellowjackets returned. These wasps were able to find solutions to the problem at hand—an entomologist threatening their nest. Their solution was simple: crawl through the mesh of the bee veil and sting. Four to five stings later I attempted to set the 100-meter-dash record in a bee suit in the downward direction toward base camp. There I was greeted with apprehension by Frank, who said, "Don't bring those wasps down here." Hayward, being more fatherly and understanding, helped me put on an army issue, green mosquito head veil under my bee veil. That should solve the problem. Wrong again. Earlier lesson learned from yellowjackets reinforced. This time the beautiful black wasps flew through the veil and summarily crawled under the elastic of the mosquito netting. Half a dozen stings later and it was déjà vu, a screaming 100-meter dash in full garb back to base camp. This time a few wasps accompanied me. As a wasp buzzed him, Frank's demeanor changed from apprehension to annoyance: "Don't come back here again. Stay away with your wasps. I don't want to get stung." Undaunted, though in pain—these wasps hurt a lot more than yellowjackets or honey bees—and with Hayward's help, we tried again. This time Hayward liberally applied silver duct tape around the entire juncture between the mosquito veil and my sweatshirt under the suit and around the juncture where my pants covered my boots and my sleeves covered my surgeon's nitrile gloves. Ignoring Frank's mutterings, off up the hill I went. This time sweet success. The prize was mine, and a major gap in my data on venoms of stinging insects was filled.

Who are these yellowjackets and wasps? "Wasp" traditionally refers to any member of the hymenopteran family Vespidae, a group that includes hornets, yellowjackets, "paper wasps" in the genus *Polistes,* and an assortment of other, mostly tropical, social wasps that live in social colonies and usually make their nests of paper. The word originated from the Anglo Saxon word root "webh," meaning to weave, a reference to the "weaving of wood fibers" to make the paper of their nests. Today, in Europe "wasp" refers to hornets (*Vespa*) and their diminutive versions (*Vespula, Dolichovespula*). Oddly, in the United States, the word "yellowjacket" is commonly used instead of wasp for *Vespula* and *Dolichovespula.* As if more is needed to make one's head spin, Americans single out the largest yellowjacket species and give it the name "baldfaced" hornet. That is, unless one is an American physician, in which case the favored name for the species is "white faced" hornet. All this seems rather silly, given that the species (*Dolichovespula maculata*) is neither a hornet (any wasp in the genus *Vespa*) nor does it have either a "bald" (whatever that might mean here) or a "white" face. Americans and American physicians are not the only people to add naming confusion. Scientists have contributed their part. The most abundant yellowjacket species in most of the western United States, especially west of the Rocky Mountains, was named in 1857 by the Swiss-born Henri de Saussure as *Vespula pensylvania.* He got it wrong on two fronts. First, Pennsylvania was a large well-recognized United States eastern state, and the name was universally associated with that part of the country. Second, he misspelled "Pennsylvania," leaving out an "n." The second mistake could be forgiven, as de Saussure was neither an American nor an Anglophone, and he simply might have made a typographic error, or have been unaware of the correct spelling of William Penn or Pennsylvania. The first mistake is harder to forgive, as de Saussure confused combined species taken from eastern Canada and from western Mexico, giving the mixture of species the name "pensylvania." Neither set of specimens came from Pennsylvania, leading one to wonder where the pennsylvania (or pensylvania) came from. To add more flavor to the confusion, Joseph Bequaert in

1931, while attempting to clarify the mess listed the name as "pensylvaniva" ("v" emphasized), clearly a typographic error, but one permanently in the record.[2] The net result is that to this day the name causes both spelling and distribution problems in the literature. No wonder the best name for the species might be its official common name: the western yellowjacket. In the spirit of facilitating communication, this writer will use the word "yellowjacket" to specify the combined genera *Vespula* and *Dolichovespula*, "hornet" to specify *Vespa*, and "wasp" for any social wasp, including yellowjackets and hornets.

Yellowjackets and hornets are large, often shiny insects sporting coats of yellow or white on black, sometimes splashed with reds, oranges, or browns. They mostly have an annual life cycle in which a single individual, a fertilized queen, founds a colony alone. Later in the cycle the queen becomes an egg-laying machine, leaving her offspring, the workers, to do most of the work. The cycle starts with males and young queens that usually fly out of the colony to mate. Depending on the species, mating is a short-lived activity, lasting anywhere from 10 seconds to 10 minutes, which looks to the human observer as awkward, rather than the frenzied sex of fire ant matings. The male mounts the female from behind, engages genital fixation, and then falls backward, often dangling from the female. Both males and females mate multiple times, averaging over five and nine times for two studied species.[3]

Once mating is complete, the new queens fatten themselves, while the not-so-lucky males die (being a male insect can be tough). In temperate-climate species, each female then locates a protected area, often under tree bark, within decomposing forest litter, or within crevices in buildings to "hibernate" through the winter. Months later, after using up to 85 percent of their fat stores,[4] queens emerge from overwintering and start new colonies. The queen's first goal is to find a preferred nest site: an abandoned rodent burrow, some other hole in the ground, a nice location in vegetation, a hollow tree, or a space in the wall of a home. Once the queen locates her nest, she makes hexagonal paper honeycomb cells from chewed wood or plant fiber, lays an egg in each cell, and surrounds the whole with a paper envelope. When eggs hatch, she forages from her nest to obtain prey to feed the larvae. As the young grow, the queen will often curl around the thin pillar-like pedicel from which the nest is attached and warm the young to speed their maturation. If all goes well, in a few weeks, young workers will emerge, take over most of the work duties of foraging, collecting pulp and water, and enlarging the nest, and bring the colony into the rapid growth phase.

Often, all is not well. A queen may covet the successful nest of another queen. Why make your own nest if you can steal someone else's? Other queens will try to invade and take over (usurp) the host queen's colony, often killing the queen in the process. These usurping queens can belong to the same species or to another species.

One well-studied example is the eastern yellowjacket, *Vespula maculifrons,* and the southern yellowjacket, *V. squamosa,* in which the larger southern yellowjacket queen prefers to take over a nest of her eastern sister. Not that she can't make her own colony—she can; however, she prefers to pilfer the work of another. These usurpation battles can be violent. Often a series of separate invasions result in multiple dead queens, stung to death and found under the nest or in the entrance. Odds for survival can worsen for the new queen. She can be invaded by another species, a full social parasite, that is unable to found her own nest or even to produce workers. These cuckoos have a real advantage over the original queen. They are usually stronger, with harder body integuments and larger, more curved, stouter stingers. Their venom is not more powerful,[5] but their stings are better able to find their mark on the host queen. The host queen usually fatally loses.

If the queen makes it through the colony founding stage, and 90 percent don't,[6] the colony enters the growth phase. Food, fiber, and foraging take center stage. The queen's newly reared and reproductively sterile workers now forage distances within 400 meters from the colony, sometimes up to 1,000 meters,[7] where they seek their resources: water, nectar, fiber, or prey. Water is needed to manufacture paper or to cool off in hot weather. Nectar from flowers, honeydew sources, fruit, or soft drinks ("Coca-Cola wasps") is needed to fuel their energetic flight or for warming the nest. Fiber is needed to enlarge the nest by adding more paper cells or protective envelope coverings. Different species have different favorite fiber sources. Some prefer weathered sound wood, as from the gray burnish on the clothespins in my backyard, others prefer rotting, punky fiber sources. The latter paper produces fragile, trashy paper that easily crumbles, much to the chagrin of collectors hoping to attach a wasp nest trophy to their wall. Prey foraging is the most difficult task for a worker. She must first locate a suitable prey or other protein source, such as carrion. Then she must catch it (carrion excepted), subdue it, process it to be suitable to carry, and fly back with her prized meatball to the nest. Favorite prey include flies, especially houseflies, stable flies, horseflies, or other common flies and caterpillars, but most any insect or spider prey will do. The list reads like an inventory of small life: moths, grasshoppers, cockroaches, cicadas, beetle grubs, bees, spiders, and even other yellowjackets of the same species.[8,9] Large life is not excluded from the diet. Yellowjackets have been known to forage for flesh from horses' open wounds.[10] One particularly intrepid entomologist recorded notes of a yellowjacket carving a hole in his ear lobe and flying off with a drop of blood in its mouth.[11]

Wasps forage both visually and olfactorily. Their large compound eyes are better suited for detecting movement than for forming sharp images. If a prey moves, it is detected and pounced on; if the prey does not move, as in a sitting fly on a barn wall, the wasp will pounce on it anyway. Imagine the frustration of a yellowjacket repeatedly pouncing on nail heads on barn sides that look like resting flies. To its credit, a yellowjacket learns that a nail is not a fly and does not pounce on it again. It, however, pounces on other nearby nailheads, having to learn each time that this black spot is not a fly.[8] Odor is the other main cue in foraging. Yellowjackets are often seen flying upwind to a food source. If the food source is too large to be carried home in one trip, a yellowjacket will make an orientation flight in which it hovers while facing the source and arcing from side to side while progressively moving away from the location. With this method, the forager visually learns the location of the bounty, whether it is the remnant of a large spider, a dead mouse, or a partially eaten jelly sandwich, and can quickly return for its next morsel. Yellowjackets also recruit fellow nestmates to a food source by transferring the food odor to other yellowjackets in the nest. Armed with the odor cue, these yellowjackets head out and search for the odor source and visually for other yellowjackets already feeding on the source.[12]

As the yellowjacket colony grows from a small queen colony to a large, populous colony, it changes from a factory that produces only workers to one that produces reproductive males and queens in addition to essentially sterile workers. This shift usually occurs in late summer and autumn, when the colony is most populated. Like having a house full of teenagers preoccupied with the other sex, life in the yellowjacket colony is more chaotic when the new reproductives arrive. They do not work, but they demand and eat food. The colony begins its decline, often with the queen mother disappearing and the workforce shrinking. By the end of the season, all workers have died, the new queens have mated, and the males have died. So ends the annual cycle, as the nest is abandoned, and the new queens seek their refugia for the winter.

But, wait. This is not always the story. In warmer parts of their range, a few biannual colonies of some species overwinter into the next year. These polygynous colonies, with multiple functional queens, continue to grow unabated. Colonies sometimes have more than 100 queens,[13] grow to 3 meters high and 1 meter in diameter,[14] and weigh 450 kilograms.[15] Young children might be advised not to throw rocks at these colonies.

Rock-throwing children are not the only "predators" of yellowjackets. Yellowjacket predators come in all sizes. Small predators include robber flies (Asilidae), spiders, and dragonflies. Robber flies and dragonflies catch individual foraging queens, workers, or males while on the wing. Robber flies grab the forager and punch through the neck or top of the thorax with their stiletto mouthparts and inject a powerful, nearly instantaneously lethal venom. Dragonflies swoop onto a flying wasp, hold it in a basket formed with their six legs, and quickly chew through the wasp's body. Web spiders prey on yellowjackets caught in their webs, while crab spiders cryptically hide in flowers and grab individuals as they land on the flower, seeking nectar.

Large predators include a variety of birds and mammals. Mice, moles, and shrews commonly predate hibernating queens. Large mammals pique human interest more than mice and moles; they also are more serious predators of yellowjackets, often destroying fully mature and populous colonies. In Great Britain, badgers are particularly important.[15] Once, while they were excavating a yellowjacket colony in a Wisconsin backyard, Jenny Jandt and Bob Jeanne, two talented yellowjacket experts, watched a large raccoon sitting on the back porch. The raccoon, in turn, was watching them in apparent anticipation of a future meal of yellowjacket nest scraps that might be left behind. (Little did it know that Jenny and Bob would leave no scraps.) Although this raccoon was unsuccessful in getting a meal, raccoons are considered the most important predators of underground yellowjackets in eastern North American, where they avidly excavate the nests, scatter the combs, and eat the brood from the combs much like a person eating corn on the cob.[16] In Britain, other large predators include stoats and possibly weasels, whereas in North America, skunks, badgers, and black bears can be important predators in regions where these animals are abundant. Exactly how large predators endure yellowjacket stings is unclear. What is clear to humans is that yellowjacket stings really hurt. Do these animals have tough enough skin and dense enough fur to prevent stings? This seems unlikely, especially around the eyes, nose, and mouth, where skin is thin and hair is short. I once counted 3,305 honey bee stings in a German shepherd. Ninety percent of these stings were on the face, especially around the eyes and muzzle.[17] I would be surprised if stinging yellowjackets were less able than honey bees to deliver a stinging message. Bears, famous in cartoons for their love of honey, have an equal love for yellowjacket protein-rich grubs. This love apparently transcends sting pain, as described by N. K. Bigelow in 1922: "A nest in the ground they will scratch up, digging with much rapidity, but often having to stop from the stings of the enraged insects. They will snarl and roll on the ground and go at it again. Although the punishment is severe Bruin keeps at it until he has secured his hard earned prize."[18]

Perhaps bears, badgers, raccoons, and skunks are simply tougher and better at enduring stings than humans. Perhaps they are simply hungrier. Maybe they are resistant to venom, able to neutralize its effect like a mongoose neutralizes the effects of cobra venom. We still don't know the answers to these questions. Stay tuned!

Birds are also important predators of yellowjackets and other stinging insects. A variety of birds, including Eurasian blackbirds, great tits, and kingbirds, will take yellowjackets in midflight. Another group of birds is so good at predating bees and wasps on the wing that they are called bee-eaters. The European bee-eater, *Merops apiaster,* catches a flying wasp, beats it against a branch to purge its venom, and then down the hatch it goes. If it's a male wasp, the bird dispenses with the venom-removing operations and directly consumes it.[19] The peculiarly named honey buzzard (*Pernis apivorus*), a large Old World bird, specializes in eating stinging insect prey. It has a delicate beak and is not closely related to hawks (*Buteo*) and certainly not to vultures, commonly called buzzards in North America. Honey buzzards are fond of yellowjackets and other stinging insects. The bird appears casual in its excavation and consumption of larvae and pupae taken from nests. It seems unconcerned with swarms of yellowjackets surrounding its head and shows no sign of being stung. The bird seems more preoccupied with looking for its own predators and enjoying its meal.

If a variety of large, medium, and small predators prey on yellowjackets, what, then, is the value of the sting? Is the sting simply useful for killing or paralyzing prey? The brief answer to the first question is that the sting is a marvelous defense against most potential predators. The exceptions to the effectiveness of the sting as a defense alert us to how evolution is constantly honing predatory adaptations, strategies, and defensive behaviors. Exceptions also allow us to appreciate the usual value of the defense. Focusing on the success of a defense, though often less glamorous than the failures, is crucial to understanding the organism's life.

Answers to the question, Do yellowjackets sting their prey? have filled the literature with anecdotes, poor observations, and careless oversights. Common sense tells us wasps should sting prey, but common sense is a tricky thing. Quoting my high school physics teacher: "Common sense is a very uncommon thing, because so few people have it." Common sense easily biases our observations. Some classic reports that claim or suggest prey are stung include an early, particularly brash statement by F. M. Duncan in 1911: "It is chiefly as a means of procuring the necessary animal food for her young that the mother wasp uses her sting. . . . [Flies are killed by] repeated stab of the wasp's sting."[8]

Others were more cautious and concluded that the sting is used for large, powerful prey. These authors include Phil Rau: "Another [yellowjacket] was seen to sting an adult [grass]hopper, follow it in its agonized flight, and sting repeatedly until lost from view."[20]

Phillip Spradbery wrote: "Only in rare instances does the wasp use its sting when grappling with prey and then only when it is particularly large or struggles sufficiently to free itself from the wasp's grasp."[15]

When precedents based on hasty or poor observations find their way into the literature, they become impossible to eliminate. Other common themes in sagas of stung yellowjacket prey relate to dangerous prey. These include tales of yellowjackets and bees: "[The wasp] always trying for the head grip, thereby keeping the bee on its back. As soon as the bee was tiring the wasp became more aggressive, holding on with its legs and stinging it in the thorax."[21] And "after stinging it [a honey bee] several times between segments and biting it severely, the hornet carried it away to devour at leisure."[22]

Further examples involve wasps stinging dangerous opponents: a wasp entangled in a spider's web stings the spider;[23] a wasp initially captured by a dragonfly turns the tables, stinging the dragonfly and treating the stung dragonfly as prey.[24] In an observant discussion in reference to a yellowjacket capturing a fly, F. J. O'Rourke writes: "The wasp using its sting vigorously but at random," and continues with "when it got on top of the fly it used its sting more vigorously while, at the same time, it chewed the interval between the head and thorax of the fly . . . [and] killed it, not by means of the sting, but by sawing off her victim's head."[25]

Two elements of wasp behavior color prey-stinging observations. Capturing prey is an energetic endeavor that requires oxygen. Yellowjackets, like most insects, breathe via tracheal tubes to their tissues with air moved by abdominal pumping much like an accordion. Because the sting is sheathed inside the pointed tip of the abdomen and is thin and black, it is impossible to see the sting when it is ephemerally exposed during stinging. Our expectations lead us to conclude that abdominal pumping drives home the sting. Another befuddling feature of prey-capture behavior is purely mechanical. We naturally carry infants or other heavy objects by holding them against our side and supporting them on our hips. Our hip serves as an extra arm. Yellowjackets have no hips, but, like us, sometimes need an extra arm to control and manipulate struggling prey. The tip of the abdomen is that handy, mobile extra arm. This use of the abdomen fits expectations perfectly for stinging.

A final factor in our prey-stinging tale is the confusion of defense with predation. When a predator attacks, a yellowjacket or other stinging insect will attempt to sting in personal defense. Such defense was witnessed in the case of wasps struggling with attacking spiders or dragonflies. If the wasp wins the struggle, its behavior then changes to predatory and the former predator is butchered and carried off to the nest. Carl Duncan expressed prey stinging succinctly: "These statements the writer believes to be simply errors based on preconceptions rather than precise observations."[8]

Ironically, stings must sometimes fail if they are to become effective defenses. If yellowjackets had no predators, their venomous stings would never have evolved in the first place. Natural selection quickly weeds out unnecessary energy-requiring body parts. If the sting is not needed, it would have gone the way of blind cave fish eyes; the ancestors of blind cave fish had perfectly functional eyes. Alternatively, if the sting were only marginally effective against major predators and competitors, it might have gone the way of forest thatching ants, whose sting modified into a nozzle that sprays formic acid. The key to the evolution of the sting was that it sometimes worked. Those queens and their workers that possessed stings that defeated more predators than other queens and their workers with less effective stings would be at a selective advantage for passing their genes, including their particular version of the sting, to the next generation. Predator filtering of the gene pool for stings is the driving force for the evolution, improvement, and maintenance of the sting. As the yellowjacket sting became more effective against more potential predators, it opened ecological doors of opportunity. Now, rather than being defenseless tasty morsels, leading unobtrusive, restricted lives, defended yellowjackets could venture afield during daylight to visit flowers for sweets or fresh cow pats in the meadow for flies. And produce more babies.

I have experienced firsthand one additional benefit of wasp stings (for the wasp, that is). Like the forest thatching ants, some wasps also spray venom, not formic acid but proteinaceous venom with lytic and painful components. I first suspected this behavior as I was antagonizing a colony of eastern yellowjackets. I succeeded in arousing hundreds of workers to fly around my bee veil, attempting to enter and sting. Suddenly, the air became redolent with a sweet-smelling perfume. The odor was rather pleasant and flowery overall, even if the situation was less than pleasant. Where did this odor come from, and what was its function? The answer to the second question was abundantly clear. As soon as the odor appeared, the ferocity of the yellowjacket attack increased dramatically. The odor was a pheromone signaling alarm and recruiting more sisters to battle. Its source? I suspected the venom. Back in the lab when a fresh venom sac was crushed, sure enough, the odor arose. No other part of a yellowjacket produced the odor. Also, while the odor was present in the field, the air around my face became irritating and unpleasant. The workers were spraying microdroplets of venom into the air. These droplets were release devices for the pheromone.

Yellowjacket venom was not that bad when it was sprayed into the air. At least it didn't affect me directly. That was not the case with another species of social wasp in the tropics. *Parachartergus fraternus*, a lovely, delicate shiny black wasp with clearish white wing tips, makes beautiful artistic nests of undulating delicate waves that form into a thin, gray paper that covers the combs. In Costa Rica, nests are often built in small trees several meters aboveground. The mere fact that these wasps are black signals *beware*.

One day while driving along the steep road to Monteverde, Costa Rica, with my assistant, we saw a particularly nice nest on a small tree on the left side of the road. The nest was located about 3 meters up in a 15-cm-diameter tree that was leaning at an angle about 20 degrees over the abyss of the valley below. The operation seemed a snap. Simply don my bee suit, shinny up the tree with a bag in hand, carefully slip the bag around the nest, break off the branch holding the nest, and, presto, I would have the nest. The wasps seemed to have other ideas. As soon as I started climbing, the vibrations alerted them, and they watched but did not fly or attack, even when I was right next to them. During this time, I was holding my breath to prevent a massive attack. (That part of the operation worked.) All was going as planned until the bag hit a snag and failed to surround the entire nest. This effrontery was too much. The wasps exploded off the nest at me. They couldn't get through the veil, but they had another trick up their sleeves. They sprayed streams of venom through the mesh of the veil directly at my eyes. As soon as the first bit of venom hit my eyes, I closed them tight, preventing more potential damage and pain. Meanwhile, I'm up 3 meters over a cliff with a nest only partially in the bag, and I can't see. Not wanting to lose this opportunity, I somehow got the nest entirely in the bag, broke off the attaching branches, and slid blind down the tree, prize in hand. My helper led me to the car and drove us off. The pain and eye watering continued for some minutes; fortunately, the venom is water soluble and the tears eventually washed it away.

We humans do not consider yellowjackets among our best friends. In the words of Howard Evans and Mary Jane West-Eberhard, "Wasps are not the most popular of animals, for it is hard for us to reconcile ourselves to creatures so well equipped to defend themselves."[26] The theme is continued by Harry Davis: "Almost without exception, people did not want the wasps on their property, and their primary concern was being stung."[27] The earliest written record of wasps relates to Egyptian King Menes, the long-ruling, powerful first Egyptian pharaoh. Historical fancy records his death as caused by a wasp sting. As the story goes, King Menes was stung and died about 2641 BCE, while adventuring on a warship near Britain. As delightful and imaginative as wasp lovers and allergists would like this story to be, the king was not killed by a wasp. Most likely he was killed by a hippopotamus while navigating the waters of the Nile River. The inner message of this story is unclear: Is it simply to show how strongly wasps affect our emotions and fears? Or does it imply that we are more frightened of wasps than hippos?

Aristotle, some 2,300 years ago, was the first scientist to write about yellowjackets and hornets. He described their stings as stronger than the stings of honey bees. Aristotle provided many accurate descriptions of their lives. He noted that drones were stingless and debated whether their leaders (queens) had stings. (He concluded that they likely had stings but did not put them to use.) An era of superstition and romantic ignorance followed Aristotle. Romans believed that yellowjackets were generated from dead horses, that hornets were special, emanating from dead warhorses, and that honey bees emerged from dead bulls. These beliefs continued into sixteenth-century Europe. Only in 1719 were the beginnings of modern scientific understanding of wasps established by the observant French naturalist M. de Reaumur.[28]

Do these stories and history tell us that yellowjackets and hornets are winners in the human-stinging insect game of life? Several lines of evidence suggest the answer is yes. In the United States, only about 50 people die per year from the combined stings of all stinging insects (wasps of various types, honey bees, and fire ants).[29] At the same time, 10,000 times as many people die from smoking; and diabetes kills a thousand times as many. Do we tell stories at cocktail parties of our narrow escape from a smoke-filled environment or of overcoming the risk of that enticing sugary, fatty doughnut at the coffee break? But we relish telling stories about surviving an encounter with a stinging insect. The message seems clear. Stinging insects win the emotional fear game. We are afraid, if not outright petrified, of stinging insects; yet, we fear not smoking, diabetes, and other many-times-more-dangerous and preventable aspects of life. Until the advent of modern technology and ways of killing yellowjackets, the wasps won, and we largely left them alone to go their own way, doing exactly what they wanted.

The evolutionary mind game between humans and hornets has been fascinating. We do not simply fear stinging insects; we entertain ourselves with our fear of them. We relish and embellish the fearfulness of stinging insects to make an even better story. One sunny July afternoon in 1999 an editor of *Cosmopolitan* magazine called and left a message that she wished to interview me. Not fancying myself as an expert on women's social culture or fashion, I asked my student Andrea for background information about the magazine. You should have seen the horror on her face: "*Cosmo* wants to interview you?" With trepidation, I approached the interview. To my relief, *Cosmopolitan* sought my knowledge about stinging insects. Why was *Cosmopolitan* interested in yellowjackets? Turned out they were worried about the safety of their young readers as they frolicked in the woods during fine autumn afternoons, and the editors wished to write a reassuring story. Score one point for the yellowjackets.

More recently, our old friend *Vespa mandarinia* (the mandarin hornet), the largest stinging insect in the world, has been the subject of news stories in China. Headlines read: "Killer Hornets Rampage through China," "Giant Killer Hornets Kill 42 People in China and Injure More Than 1600 Others," "Giant Asian Hornets Are Killing People in China, Breeding in Large Numbers." One article pictured four hornets that spanned the entire width of a person's hand. My first thought was, "Wow, those sure are huge hornets," which presumably was the intended impression. My internal BS app kicked in. Something was awry. I just happened to have in my insect cabinet, three steps from my desk, two queens, the largest individuals of mandarin hornets, which, in turn, are the largest of all hornets. I placed these in my left hand. They only spanned slightly over half the width of my hand. My hands are average to smallish for adults. How could that be? I caught my queens in Yun Xi (pronounced "Ywen shee") bamboo forest near Hangzhou, China, where they appeared to be hummingbirds slowly cruising the forest floor. These queens were the real thing. The article didn't specify the age of the person, nor show more than the hand. I put the hornet in my 11-year-old son's hand and perfect match! Apparently, as if wasps and hornets are not already big enough, we have to make them even bigger. Score another point for yellowjackets and hornets.

The power of yellowjackets to influence human emotion extends even to our legal system. Yellowjackets fall under the English tort law section covering "knowingly and willfully harbouring a dangerous vicious animal." Flies, grasshoppers, and praying mantises don't fall into that category, but bees and wasps do. I was brought in as an expert witness in a case in which a lady purchased a strawberry cake roll from a large national retail chain store in Billings, Montana. She got the munchies near midnight one evening and ate a slice of this cake roll. No problem. However, the next morning she ate a piece of cake and was stung by a "bee" in the cake roll, had an allergic reaction, and was treated at the hospital emergency department. Was this product liability of the cake manufacturer or of the retail store? I testified that the German yellowjacket (not a bee) could not have been alive at the time of the perceived sting, as the cake was manufactured and sealed in plastic a few days earlier, 1,600 miles east, and that the yellowjacket was thoroughly embedded in pink frosting, evidence it was caught in the cake during manufacture. A worker wasp could survive that treatment, at most, for a few hours. And, no, the wasp couldn't have flown into the cake at the store because the cake was sealed at the time of purchase. Further examination revealed the sting of the gooey and somewhat crispy yellowjacket was completely withdrawn into its abdomen. The sting tip was intact, not bent, and not broken off. Case closed. The wasp was innocent, and the lawyers and the wasp won. Score another point for the yellowjackets in the psychological battle with us.

As with fire ants, humanity has not lost the battle against yellowjackets willingly. Dislike and economic damage are a prescription for war. Angry fruit growers dealt with damaged crops, park and resort operators closed or limited activities, loggers shut down operations, firefighters were thwarted as they fought forest fires, and beekeepers' beehives were under attack all because of yellowjackets.[30] Action was needed to combat the yellowjackets. Lead arsenate to the rescue. But it didn't kill colonies. Next, the miracle insecticides—DDT and chlordane—were added to horsemeat baits. Intense baiting lowered the worker populations in local areas,[31] but insecticides are environmentally harmful. Mirex, the wonder insecticide developed for fire ants, also worked in baits but was environmentally unacceptable.

An intense effort was devoted to finding the perfect yellowjacket bait. Initially, tuna fish baits and other fish flavors were favored, especially Puss'n Boots fish-flavored cat food.[32,33] The reign of cat food did not go unchallenged. John MacDonald, research professor at Purdue University, and colleagues visited a local zoo and observed that yellowjackets actively scavenged from Nebraska Brands feline food fed to the zoo's big cats, not domestic felines; nevertheless, it was highly attractive to yellowjackets. This horsemeat-based diet was vastly superior in attracting these Lilliputian carnivores than Puss'n Boots or four other tested cat foods.[34] Two years later boiled ham was found more attractive than the Nebraska Brands horsemeat product.[35] Not to be outdone, E. B. Spurr in 1995 tested nine types of fish and seven meat types for attractiveness to yellowjackets in New Zealand.[36] Venison was most preferred followed by hare and horse; beef was least preferred (all fish were between beef and horse). Should picnickers avoid venison or rabbit sandwiches in preference to beef?

In New Zealand, a country plagued by explosive populations of invasive German yellowjackets, one proposal was to pay a bounty on each yellowjacket queen captured to prevent the establishment of colonies. The hunt was enormously successful with kids (and adults) enthusiastically delivering 118,000 queens over three months. Everybody had fun in this adventure until the next season when the yellowjacket population appeared unaffected. A similar bounty program was conducted in winter in Cyprus, again with enthusiasm (huge amounts of money was paid), only to be followed the next year by one of the worst wasp seasons in years.[15] Yellowjackets were the winners again, but at least these eradication programs were environmentally benign.

Generals in the yellowjacket wars concluded that a new approach was needed. The best baits had twin disadvantages: they contained toxic materials and rapidly spoiled or became dry and unpalatable. Instead of poisoning colonies with baits that yellowjackets take back to the nest, why not deal directly with the foragers? After all, the foragers were the problem, not the colonies. Why not trap them directly and take them immediately out of the system? Thus, nontoxic chemical attractants in one-way traps became the preferred means of yellowjacket control. Harry Davis, a man of concise writing and practical approaches, tackled the problem head on. Over several years, he and his group tested innumerable compounds for attractiveness to yellowjackets, culminating with a massive screening of 293 different attractants.[37] From these screenings, first came the attractant 2,4-hexadienyl butyrate, then heptyl butyrate, and finally octyl butyrate. Over four days, these attractants lured 200,000 foraging yellowjackets to their doom in traps—enough to fill a wheelbarrow—and successfully saved the crop on an 8-hectare peach orchard. Thus, we could win small battles against yellowjackets. Perhaps that is enough.

One way to win these tiny battles with yellowjackets is to get up close and personal. If a yellowjacket stings us, usually because we are too close to a colony, we can get revenge. Locate the nest entrance somewhere nearby, typically in our yard. Over the years, many solutions have been proposed, including a variety of toxic insecticides that are puffed, sprayed, or flooded down the entrance tunnel at night, followed by sealing the entrance (be advised to use a red light, wear protection, and keep from breathing on or near the nest). Several more benign materials have also been proposed for use in place of toxic insecticides. Curiously, perhaps for legal reasons, a most effective method is almost never mentioned in American literature. Every rural or farm child knows that gasoline (or kerosene) kills insects instantly. Gasoline is not registered for control of yellowjackets or any other insect, so professionals cannot recommend or even acknowledge its use. I am not recommending it either, just providing some history of this fraught-with-danger folk remedy.

I experienced firsthand yellowjacket control with gasoline while making fire trails for a logging company in the Pacific Northwest. We cut and removed vegetation and scraped to bare earth a trail 2 meters wide around an area to be logged. That way, if a spark from a piece of machinery ignited a fire, we had a barrier and could stop the fire at the fire trail. The team consisted of four people: two buckers with chain saws to cut through downed logs and saplings in the way and two people to wield hoe dads (oversized mattocks) to scrape to the earth. All too frequently, the buckers stirred up a yellowjacket nest. Instant command: "Grab the gasoline can." Someone would grab the goosenecked gas can, carried along to refill the chain saws, stick the spout down the nest entrance, and pour in a liberal amount of gasoline. Problem solved. A few expletives later and we were back on the job. Fire and gasoline do not mix; they explode. The buckers never lit the gasoline. After all, we were preventing, not making fires. But burning is an unfortunate human national pastime. We burn fields, roadside vegetation, and yard waste. Fire is fascinating. And by human nature burning yellowjackets seems especially exciting and satisfying. The average person all too often lights the gasoline after pouring it down the yellowjacket nest. Never mind that this is dangerous and ineffective because all it accomplishes is making fire aboveground where there are no yellowjackets and fire more quickly evaporates the gasoline inside the nest that is actually doing the killing. Word to the wise: don't light it. Doing so might rip disaster from the jaws of victory in a battle with yellowjackets. Another, even sillier idea was proposed in 1770 for colony control. Add "wetted" gunpowder, light it, and the fumes will kill the yellowjackets.[38] And we think modern-day people live dangerously.

"Love thy enemy" we are advised. If yellowjackets are our enemy, as our words and actions suggest, should we also not love them? The underlying theme advises that within an enemy is inherent good. What good are yellowjackets, other than providing material for our thrills, excitement, and stories? But, wait, yellowjackets can be fine friends, albeit not fuzzy, touchy friends. Two favorite yellowjacket foods are flies, including biting and disease-carrying flies, and crop-loving caterpillars. Children who live on farms have been entertained watching baldfaced hornets and yellowjackets pounce on barn flies resting on weathered wood. Bryson expressed the enormous toll on flies 150 years ago:

> the practical result of destroying all the wasps on Sir T. Brisbane's estate was, that in two years' time the place was infested, like Egypt, with a plague of flies. . . . We do not readily appreciate the indirect benefits which we derive from the labours of wasps, just as we are not perhaps properly grateful to beasts of prey for their equally unsolicited assistance. Cats and weasels, and foxes, though they are not good to eat, are often much more acceptable neighbours to the farmer than rabbits.[39]

If one were reincarnated as a Paraguayan cow, or the rancher of the cow, another social wasp, *Polybia occidentalis,* would be a friend, indeed, because they catch prodigious numbers of biting flies, particularly from around the eyes of cows.[26]

Caterpillars are ecological eating machines, stuffing chewed leaves into their sausage bodies at a seemingly nonstop rate. This might be fine as long as you are not the plant or the plant is not a crop. *Polistes* wasps feast nearly exclusively on caterpillars, a feature that endears them to some farmers. North Carolina tobacco farmers unwillingly share much of their tobacco leaf crop with tobacco hornworms, enormous caterpillars that relish the nicotine, grow rapidly, molt into fighter jet–shaped hawkmoths, and recently have become the darling laboratory models for insect physiology and neurobiology. These half-ounce caterpillars easily eat a dozen times their weight in juicy, prime tobacco leaves. North Carolina entomologists made little wooden shelters for the wasps, moved them into the vicinity of tobacco fields, brought about substantial reduction in numbers of hornworms, and even prevented economic loss of leaves.[40] Tobacco lovers, think twice before harming a paper wasp.

This brings us to the sting of yellowjackets and the baldfaced hornet. These stings are definitely more serious and attention grabbing than those of fire ants. They are bad. On the pain scale, both yellowjacket and baldfaced hornet stings muster the respectable pain level of 2, equal in comparison to the honey bee. Surprisingly, the much larger and scarier baldfaced hornet actually hurts the same or slightly less than a yellowjacket, perhaps indicating that baldfaced hornets are better at the intimidation game. In any event, the stings of either produce instantaneous, hot, burning, complex pain that gets one's attention no matter what other thoughts were preoccupying the mind. The pain lasts unabated for about 2 minutes, after which it decreases gradually over the next couple of minutes, leaving us with a hot, red, enduring flare to remind us of the event in case our memory should fade. These stings are worthy of storytelling to loved ones.

8

HARVESTER ANTS

Pogonomyrmex californicus *is undoubtedly the fiercest, the boldest and the most irascible ant of the Sonoran Desert. Furthermore it is the quickest to sting and the effects of the sting are the most painful.* —George C. and Jeanette Wheeler, *Ants of Deep Canyon*, 1973

William Steel Creighton wrote in 1950, "There is a persistent belief that in the days when the West was wilder than it is now, Indians would sometimes stake out a human victim across a nest of *Pogonomyrmex* [harvester ants]. If this was actually done, it would be hard to imagine a more excruciating death."[1] Years earlier, William Morton Wheeler in his classic 1910 book *Ants* sounded a similar story: "If it be true, as has been reported that the ancient Mexicans tortured or even killed their enemies by binding them to ant-nests, *P[ogonomyrmex] barbatus* was certainly the species employed in this atrocious practice."[2] Whether these stories have any veracity or are just urban legends is hard to know, though Jeffery Lockwood in his 2009 book *Six-Legged Soldiers* provides evidence that there might be some truth to the concept of these stories.[3] Men of the Northern Miwok in central California voluntarily stood or laid on disturbed harvester ant nests to determine who of the four or five men was strongest. The chief awarded a prize to the man who lasted the longest on the ant mound.[4] Ever since literate people met them, harvester ants have attracted immense popular interest and imagination. These ants not only make impressive and conspicuous nests and populate children's ant farms but also deliver the most painful and unusual stings of any North American stinging insects.

One might be tempted to generalize from experience with stings from honey bees, yellowjackets, baldfaced hornets, various wasps, bumble bees, sweat bees, and even fire ants that all insect stings are sort of like a bee sting, differing mainly in intensity. Anyone who has been stung by a harvester ant knows better. Harvester ants are docile giants of the temperate ant world, unobtrusively going about their business of harvesting seeds for food. They have no attitude like fire ants and cause no harm if left alone. If they are sat on or pinched, however, they deliver a sting that is nothing like a bee sting. The pain is intense, comes in waves, and is deeply visceral. The intense pain lasts 4 to 8 hours, not 4 to 8 minutes, as with a typical honey bee sting.

In addition to *Pogonomyrmex*, a variety of other ants collect seeds, including *Messor*, the famed ant of biblical times; *Pheidole*, the world's largest genus of ants; *Aphaenogaster cockerelli*, the long-legged desert ant; and some fire ants. *Pogonomyrmex* harvester ants, often considered the true harvester ants, are the most conspicuous of the seed-collecting ants in most areas and have captivated my imagination and that of innumerable others. Curiously, of the aforementioned ants, only the harvester and fire ants sting; the rest have nonfunctional stings. Why those other highly successful ants lost their stinging ability is unclear. A leading suggestion argues the loss is a result of competition and predation by other ants, against which the sting is a poor weapon. Fire ants circumvented the problem of an ineffective sting by having an unusual and highly effective venom that when daubed or sprayed on other ants penetrates their waxy protective integumental barrier and kills or disables them.[5] Sadly, for the harvester ants, their venom is harmless when topically applied to other ants. Given this venom handicap, these big, slow, methodical, unassuming ants of the American deserts and southeast are doing well. Their name suggests a life of hard work and no play, projecting an image of boredom and a boring life. But boring can be brilliant. Harvester ants have brilliantly carved an enormous niche for themselves in nature and etched indelible images in mankind: images from cartoon depictions of a backyard volcano erupting with angry ants when their home is trespassed, to images of vast stretches of the American West marred with ant castles surrounded by denuded moats of dirt. Grasslands flowing from horizon to horizon, bearing scars of harvester ant mounds, scream images of Earth infected with its version of smallpox. Should we be surprised then that men declared sagebrush wars against entire communities of harvester ants? And compared to the destructive intent of the masters of these wars, can we fault young children, spying big red harvester ants and big black harvester ants, for having the urge to drop a red ant into the entrance of a black ant nest?

As Christmas wrapping paper sales tell of the success of the upcoming Christmas shopping season, insect common names tell of our perceptions of insects. Insect common names are so important that the Entomological Society of America maintains an official registry of insect common names. The organization, some 7,000 insect scientists, has a permanent committee whose sole job is to research, name, and oversee common names—it's that important. Common names roughly represent human interest in a given insect species and range from the obscure sugarbeet root aphid and the chicken dung fly to the common honey bee. In the battle for the most common names among ants, the prize goes to the harvester ants, which eked out victory over the second-place fire ants by having six named species compared with five for fire ants. Both groups trounced the bigheaded ant, the sole represented member of the largest genus among all ants. The number of named harvester ant species suggests that people have a driving interest in them. Even the notorious yellowjackets in the genus *Vespula* only managed four common names. Harvester ants have more common named species than any other genus of stinging insects, except bumble bees with their whopping 36 common names (apparently, Americans have a love affair with bumble bees). The common name filter yields such pleasant harvester ant names as the California, the Florida, the Maricopa, the red, the rough, and the western harvester ant.

In contrast, the original taxonomists, dealing directly with the ants, frequently chose more powerful names, often celebrating native American Indian tribes, including the Apaches, Comanches, Maricopas, and Pimas, or scientific names, including *desertorum, bigbendensis, huachucanus,* and *anzensis* (referring to the Anza Borrego Desert of California), signifying the harsh environments some species inhabit. Some names were simply poor choices, as exemplified by *Pogonomyrmex bicolor,* a reference to the red anterior and black posterior of the ants originally discovered and described. When on a trip to Los Ojitos, Mexico, my wife and I found ants of this species that were essentially entirely reddish. So much for names based on color! Finally, the scientific name of perhaps the most celebrated harvester ant, *Pogonomyrmex barbatus,* translates as "the bearded beardy ant." No wonder its common name is the "red harvester ant."

Harvester ants are ant icons of the New World. The 60-odd species span from the western three provinces of Canada through the United States, Mexico, and Guatemala, skipping the rest of Central America, and picking up in South America, where they are present in every country (excepting the small northern countries of Suriname and French Guiana) to southern parts of Argentina and Chile. They have even crossed the Caribbean Sea to Hispaniola. Some harvester ants sport bright-red coats, some various shades of brown or yellow, and others black. Most are large, often 8 mm long (⅓ inch), with the largest reaching 13 mm. The life cycle of all starts the same, beginning with virgin females and males. They usually fly in a mass exodus from their maternal colonies and form mating swarms in which both sexes engage in brief, frenzied mating orgies. Multiple mating by both males and females is the general rule though exceptions occur, and, in some species, mating might occur at, or within, the colony.

An unusual grouping of harvester ants, residing in an area of the sparsely populated American West near the border between Arizona and New Mexico, take the battle of the sexes to a new high. The participants in this mating ritual are a mixture of males and females of the rough harvester ants (*Pogonomyrmex rugosus*) and the red harvester ant.

For their mating system to operate, females of each species must mate within a brief couple of hours with males of both their own species *and* males of the other species. If a female mates only with one type of male, her future is bleak. If she mates only with her own species, she can produce only reproductives, and, with no worker force, the incipient colony withers and dies. If she mates only with males of the other species, she can produce only workers and not queens. This limits her reproductive ability to producing only sexual males, a booby prize obtained by laying unfertilized eggs that turn into males. The queen's interest is to mate mainly with males from the other species and only one or two males of her own species. This way she has ample sperm from the other species' males to produce plenty of essential workers and some sperm from her own males to produce new females. The male's interest is dramatically different. If he mates with females of the other species, his sperm are wasted on nonreproductive workers, and his genetic lineage dies from lack of production of a new generation of females that carry his genes. If the male mates with females of his own species, he successfully produces daughters to carry on his lineage. Thus, conflicting battle lines are drawn between males and females: females wish to mate with lots of males of the other species; males wish to mate only with females of their own species. One problem occurs. In the short time of the frenzied mating swarm, males and females either cannot, or do not, discriminate between members of their own or the other species. Once they are engaged in the actual mating act they can distinguish. What's a male, or a female, to do now while engaged in this blissful act? If a female is mating with her own male, she terminates mating more quickly than if she is mating with the male of the other species. If the male is mating with his own female, he jacks up the speed of sperm transfer relative to the transfer of sperm to a female of the other species. The net outcome of this battle of the sexes is a wash—each partially gets what she or he wants—perhaps a good thing for both, given the alternative of a collapse in their colony population if either succeeds completely.[6]

The newly mated females, now queens, quickly get on with the business of attempting to survive and found a colony. Meanwhile, the males might remain around the mating arena for a day or two before dying. New queens, as in most ant species, have the fascinating behavior of breaking off their own wings soon after mating and beginning their new life. Males do not, and cannot, break off their wings. Queen wings are structured slightly differently from males and have preweakened areas near their base that allow the wing to snap when bent downward just the right way by the queen. Imagine the engineering necessary to allow queens to flap their wings vigorously enough to carry them upward through the air without breaking, and yet to be easily broken when the queen desires.

The wingless new queens face the most crucial and dangerous time of their lives. They need to find a new nest site quickly and to dig a tunnel to construct a protective chamber at the bottom—all before becoming a meal for someone else or becoming toast in the hot sun. In most cases, each queen ant is on her own in this task. The California harvester ant is an exception. Here, several queens often combine their efforts to make a new nest that they share together. This odd multiple-queen founding by California harvester ants is apparently an adaptation to the exceedingly harsh environment and competition faced by these queens.[7] Once the nest tunnel and chamber are completed, the queen seals the tunnel and begins the claustral colony phase in which she raises the first generation of her family. She does this by laying a few eggs and feeding her newly hatched tiny larvae from her own body reserves. Before flying on her mating flight, the queen gluttonously feasts, storing large quantities of energy-rich fat, often more than 40 percent of her total dry body weight.[7] Remember the wings she used to have? Those wings were powered by massive thoracic muscles—muscles no longer needed by a flightless queen. Between the breakdown protein from the wing muscles and the fat and protein stored in her body, the queen manages to rear 10 to 12 tiny workers. During this time, she never leaves the protective chamber she built. The one exception is, again, the California harvester ant. One hapless queen is forced to leave the nest to forage for food for the growing larvae.

Once the tiny minim workers emerge and their bodies harden, they take over colony duties from the queen, except egg laying and some pheromone production. The minims open the sealed nest tunnel to the outside and begin foraging for food. They also enlarge the nest, by digging downward and adding new chambers for the queen, brood, and stored food. This first generation of minims is short lived but, assuming all goes well, manages to raise the second generation of now full-sized ants. At the end of the first year, the colony is small, consisting of relatively few workers and the queen. During the second and third years, the colony grows rapidly, both in population size and in nest size. Usually, about the fourth year, the colony is mature and begins rearing males and females to continue the life cycle.[8]

Ask elementary school children to name the longest-lived animals and typical answers are "turtles" and "sharks." Ask them about the longest-lived insects, and knowledgeable children will likely say, "queen termites." Human interest in longevity begins early. Kids and entomologists are naturally interested in long-lived insects, including ants.

What are the longest-lived ants? Harvester ants currently appear to win the award, outlasting all other ants, including the second-longest-lived ants, the honeypot ants, desert dwellers that store honey for lean times in grape-sized individuals who form living larders within the colony. Some harvester ant colonies have been in the same location for decades and are rather obvious. The longevity of a harvester ant colony has been exasperatingly difficult to pin down. Answers are all over the place, from an average of 15 years or 17 years for a colony reared in the lab, to 22 to 43 years and even up to 29 to 58 years.[9]

A legendary story relating to harvester ant longevity began in 1942 with a publication by Charles Michener, the famed bee scientist who revolutionized the modern biology of bees and continued that revolution in 2015 at age 96. Michener began his publishing career at age 16 with broad interests in insects, including ants, and only turned to bees at the ripe old age of 26. In his 1942 article, Michener described in detail the history and behavior of a California harvester ant colony in his backyard that he had been watching since age 6. He observed the colony for 16 years, at which time it succumbed to attack from Argentine ants, bad winter weather, or both. Mich, thus, documented a longevity of 16 years.[10] In a famous, final small-print footnote, he mentioned "one [colony] of which, according to the report of the owner of the property on which it was situated was at least forty years old." Several subsequent authors focused not on Michener's factual 16-year number, but on the hearsay of a neighbor as evidence that harvester ants can live 40 years. What then is a realistic number for the longevity of a harvester ant colony, that is, of the queen harvester ant? The answer is still unclear, but our best estimate comes from detailed studies by Kathleen Keeler of Nebraska who studied 56 mounds of the western harvester ant over a 15-year period. The western harvester ant is the species suspected to live the longest of all harvester ants. For her 56 colonies, she calculated that the last colony would live to 44.9 years of age.[9] That is the best answer we have: any takers for a longer-term study? Even Keeler's study still begs the question of what causes harvester ant colonies to die of old age. Is it because the queen runs out of sperm from her original and only mating flight, which causes the demise of fire ant colonies? Does her body simply wear out? Or is it some other cause? The tools for answering those questions are available for future studies, but for now, we are left with the conclusion that harvester ant colonies live a long time.

To live up to 45 years, a queen harvester ant must remain amazingly safe and secure. How does she accomplish this? A surrounding mass of nasty stinging and biting workers is a start. Further, she never leaves the nest alone, and if she does leave, say, because the nest becomes uninhabitable due to flooding or shade, she leaves amid a cohort of workers as they move from the old colony location to the new preconstructed nest. These defenses pale in comparison to her best defense: simply hide deeply within her fortressed castle of a nest.

In my graduate student days, I needed live colonies of harvester ants to compare ants from the extreme edges of their range with others within the central part of the range. The westernmost population of the Florida harvester ant lives isolated in the small town of Amite in eastern Louisiana, home of the National Football League linebacker Rusty Chambers. Two fellow graduate students joined the challenge of excavating an entire mature colony, collecting all ants, including the queen. Fortunately, the soil in Amite is nearly pure sand, perfect for digging. One person would dig a shovel load of dirt with ants and pour it on the nearby ground. The other two would collect the ants with aspirators, those indispensable tools for myrmecologists that consist of a bottle with an incoming copper tube to suck up ants and an outgoing screened tube connected to a rubber tube to the mouth. Aspirator use is an acquired skill. One sucks just hard enough to lift the ant and pull it into the bottle, but not so hard as to bring a batch of dirt along with it. For precaution, the user learns to direct the stream of air from the rubber tube onto one's tongue. That way the dirt sticks to the tongue (rather than going to the throat or lungs) and can be conveniently spat out. (Aspirating ants is not an ideal activity in the presence of fine company.) After a few shovelfuls of dirt are removed, a cone of sand forms, allowing the ants to be exposed and readily collected as the ants and sand tumble down the sand volcano. Several hours and thousands of ants later the hole was 6 feet deep and 3 feet in diameter. Still no queen but lots of workers present. By now, a normal shovel is useless and that other essential myrmecological tool, the army shovel, becomes the tool of choice. Real army shovels, those of World War II vintage, are coveted not only for their sturdiness but also because they can be locked into a position with the shovel blade at a 90-degree angle to the handle. One person squats in the hole, digs a full load of sand, and, holding it perched like a food tray, hands the army shovel up to the aboveground people to grab by the handle. Eight feet down we found the queen with the last workers. That seems to be a safe place for the queen. I seriously doubt, even if they lived in Louisiana, that an aardvark would bother digging that deep.

We wore shorts and a light shirt, hardly armor against stings. Nevertheless, we received only three stings, a low price to pay for a full colony of harvester ants. At least that day, the Florida harvester ants were showing their Southern hospitality.

Armed with one success we headed to Lucky, Louisiana, in the northwestern part of the state and the easternmost location of the Comanche harvester ant. Lucky, a village with fewer than 300 inhabitants, is best known for notable resident Joslyn Pennywell, a model and contestant on *America's Next Top Model.* We didn't run into Joslyn, but we did find the ants. The digging was similar, and the queen was located at a depth just under 8 feet. We were having such a good time, nobody remembers whether we got stung, and, as a parting gesture of goodwill, we tossed a couple of abandoned tires into the bottom of the hole and filled it with sand; thereby preventing small children from falling in and also providing two fewer tire-breeding sites for mosquitoes.

As we demonstrated the hard way, harvester ants are famous for the depth of their nests, which rank among the deepest of ant nests. Determining harvester ant nest depths is not an easy task. Most nests are not in pleasant sand but instead are in dry, hard rocky soil. Bob Lavigne at the University of Wyoming found an innovative solution to the hard soil and depth of harvester ant colonies. He brought in a backhoe to excavate 33 nests.[11] The human marathon excavators Bill and Emma MacKay, as part of Bill's Ph.D. dissertation for the University of California at Riverside, took the effort a step further. They totally excavated by hand 126 harvester ant colonies, the record having a total depth of 4 meters, with the queen located at 3.7 meters (12.1 feet).[12] This may well be the record depth of a colony, although H. C. McCook in 1907 wrote of one nest "fortunately exposed by a deep cutting, the galleries and chambers were traced to a depth of fifteen feet."[13]

Extremes within biological systems provide ideal investigative opportunities because extraordinary features are the result of extraordinary adaptations and behaviors. Over the years, several explanations for the extreme depths of harvester ant colonies have been proposed: protection from freezing or broiling heat, protection from wildfires, protection from dryness, and protection from predators. Protection from freezing is unlikely for two reasons. First, many areas, including the southwestern deserts, Mexico, and Louisiana or Florida, have mild winters with freezing temperatures that penetrate into the soil no more than a few centimeters, yet colony depths reach at least 2 meters. Second, even in the coldest parts of their range, including in the 1,600-meter-high grasslands around Casper, Wyoming, the soil never freezes below 60 centimeters; again the harvester ant colonies reach a depth of more than 2 meters, much deeper than necessary to avoid freezing. Protection from the fierce summertime sun and surface heat is also implausible for the extreme colony depth. I have measured soil temperature at various depths in open sandy loam in Willcox, Arizona, for many years and have never recorded temperatures above 32°C at a depth of 30 centimeters, a temperature well below the lethal temperature of at least 40°C. Likewise, extreme colony depth as protection against fire seems unnecessary. Soil is an excellent thermal insulator and will prevent heating to a lethal temperature below a few centimeters, unless a flaming dead tree falls and burns directly on top of a nest. Even then, lethal temperatures would fall short of 2 meters. One study revealed that fire can actually benefit rough harvester ants by providing a supply of toasted dead insects to supplement the usual diet.[14]

Protection from desiccation and defense against predators are the two remaining explanations for the extreme colony depths of harvester ants. These roles are not mutually exclusive. Likely, both factors come into play. Clues come from other extreme desert-dwelling ants. Both the Mexican honey pot ant and *Veromessor pergandei*, another seed-collecting ant related to the famous biblical ants, construct deep nests; the former to at least 4 meters depth and the latter to greater than 3.4 meters. All three species share features of living in hot, dry desert areas and having very deep nests. Two of the three species, the harvester ants and the Mexican honey pot ants, also have extreme queen longevities. The life span of the third, *Veromessor*, is unknown, but it would be unsurprising if its life were long. Soil moisture generally increases with depth. These features support the observation that extreme depth serves the dual purposes of queen protection from predators and protection of long-lived species from the vagaries of seasonal and yearly variation in dryness.

Harvester ants, as their name suggests, collect seeds. They are good at finding seeds. Even in parched, wind-swept deserts where the human eye sees mainly bare ground and a few shrubby plants, the grass long since eaten by hungry cows, harvester ants sally forth huge distances in a quest for seeds. Some species engineer long, broad trails that can extend 30 meters or so from the nest entrance. These ant autobahns function much the same as human highways to speed transport of goods and to reduce bumps and potholes in the path. On the outward trip from the nest, a forager can race along the smooth surface like a sports car. On the return trip, an ant burdened with a seed, sometimes several times her own weight, is like a tractor-trailer truck, lumbering along much better without obstacles in the way.

A name sometimes clouds reality. The name "harvester ant" flashes mental images of a peaceful, dedicated vegetarian ant farmer methodically harvesting its grain. Subconsciously, we tend to accept as the norm this view of the "agricultural ant," so named in the popular 1879 book on the red harvester ant by H. C. McCook.[15] Other behaviors are viewed as odd exceptions not worthy of more than tucking into a back corner of the brain. In reality, harvester ants can be active predators as well as scavengers, capitalizing on opportunities to collect dead insects or parts. During much of the year in arid regions, few insects are available; consequently, harvester ants focus on finding seeds, and that is the activity we see. When the summer rains arrive, providing a bounty of insects, harvester ant foraging switches to aggressive predation. Foraging ants tend to abandon the beaten path of their trunk trails to forage in the intervening spaces. Insects encountered are attacked, and small insects are carried back to the nest. If the insect is huge, for example, a caterpillar hundreds of times larger than an ant, teams of ants gang up to attempt to subdue the prey and drag it back to the nest. In the summer, western harvester ants exhibit two personalities. During the day, they act mostly as traditional foragers for seeds. During the night, they turn into fierce predators, focusing more on finding insect prey. These disparate behaviors make sense: During the day, few insects are on the ground where temperatures reach 40°C to 60°C or more, and most insects present on the ground are other ants. During the night, temperatures are cooler, and many insects are now present on the ground surface.

In the Sonoran Desert of southwestern North America, the first major summer monsoon rain brings great excitement to the parched inhabitants—both wild animal and human. Insect life explodes. Beetles fly, many ants send forth massive swarms of reproductive alate females and males, arachnid and insect predators emerge from their hidden refugia, and termites swarm. When termites swarm, all other bets in the lives of predators are off. Winged male and female termites become the focus of all species with even an inkling of predation. Some birds swoop the air for flying termites, others land to peck ground-running termites, lizards patrol the ground eating any termite that moves, spiders pounce, and ants of all varieties pour from their nests in feeding frenzies. Harvester ants appear to be insomniac as they forage continuously day and night. Why are termites so special? To harvester ants, termites are simply mobile seeds loaded with juicy fat and protein, and these "seeds" are soft and easy to eat, unlike the usual hard, dry seeds they must normally gnaw. Termites are about the same size as a large seed, are abundant, and are perfect for harvester ants adapted for seed harvesting.

The first monsoon rain is exciting and perilous. I, like other entomologists, love this time and drop whatever else I am doing, sometimes to the displeasure of my wife, and head to the field. The presence of rattlesnakes activated in response to the rains and the resultant increased activity of rodents present one peril. The harvester ants are no less a peril. Normally, I walk in sandals unimpeded in the desert, stopping to inspect ants and colonies, rarely worrying about stings. The greatest risk is a wandering ant climbing onto my foot and inadvertently getting trapped between my sandal and the sole of my foot. My tender feet lack protective callouses—ouch. Otherwise, harvester ants frequently crawl on my bare feet and almost never attempt to sting. During the rains, this changes. Harvester ants are now omnipresent on the soil surface. They move at a faster clip than usual and readily climb on most anything, such as my foot. If they encounter something that appears to be an animal, they bite and sting. No matter how vigilant I become, inevitably I am stung. That is a price for having a love for harvester ants.

Their habit of collecting seeds has strangely led to an alleged competition between harvester ants and humankind. From the nineteenth century into the late twentieth century, people who grazed livestock on western lands perceived a low amount of grass forage to be a result of harvester ants collecting so many seeds that too few were left for producing a good crop of grasses. Add to this the emotional response of numerous anthills "marring" the aesthetics of the landscape, and we have a prescription for war against harvester ants. Lost in the fray were data on the basic biology of harvester ants. Just what percentage of the seed crop is collected by harvester ants? Far from the predictions of woe, a surprisingly low percentage is actually harvested. Estimates range from a wildly exuberant 10 percent of the seed crop to a more accurate estimate of 2 percent, a figure based on a carefully controlled experiment involving marked seeds.[16] Even using the highest estimate of seed removal, the impact of seed harvesting is minuscule compared with the damage to the grasslands caused by overgrazing and subsequent soil erosion. Opponents of harvester ants, not content with basing justification for harvester ant wars simply on seed loss, alleged serious damage from the clearing of vegetation around ant colonies, removal of young crop plants, attacking horses and livestock, stinging agricultural workers, and, worst of all, damaging farm mowers and other equipment when they hit the elevated ant mounds. Toss in damage to airplane runways and roads from undermining their surfaces and an ironclad case against harvester ants was formulated. Pesky little benefits of harvester ants, like soil aeration, nitrogen, and phosphorus enrichment of the soil around the nests, and the increased plant lushness near ant mounds can be ignored. Also forgotten were the joys of little and not-so-little children playing with horny toads, which are highly dependent on harvester ants as their main food source.

The long-established herpetologist George Knowlton of Utah State University wrote, "[Harvester] Ants are responsible for keeping thousands of acres of western range land bare of vegetation, land which otherwise would support forage plants and livestock" and anointed the harvester ant "the most economically important ant pest in Utah."[17] Initial efforts to eliminate harvester ants commenced years before modern insecticides. These early efforts were limited to powerfully toxic, nonspecific poisons. A particularly detailed study described in 1908 showed that the best killing agent was carbon disulfide, a volatile, flammable liquid that forms a vapor two and a half times heavier than air. Carbon disulfide worked best because its density caused the fumes to penetrate deeply into colonies to kill queens and inhabitants. The authors went on to detail how hydrogen cyanide gas was useless because it, being lighter than air, would not sink into mounds to kill queens or most workers and that gasoline and kerosene were even less effective, again, because they did not penetrate far into the mounds. After sealing the evidence for carbon disulfide, their final piece of advice was not to light it because "the explosion which followed did not . . . blow the vapor far down into the chambers."[18]

Not wanting to be left out, other control enthusiasts hauled out the materials of last (or was it first?) resort, various arsenic compounds, including London purple, a by-product of the garment-dyeing industry. These heavies, a step backward from carbon disulfide, failed pathetically. Soon thereafter, industrial chemists created miracle insecticides—DDT, chlordane, aldrin, dieldrin, heptachlor, toxaphene (you name it)—certain they would work. Again, the depth of the colony often prevented these insecticides from reaching the queen. Never fear, a page from the fire ant battle manual should do the trick. Enter Mirex, Kepone, and Amdro. Amdro is still used to kill harvester ants. At last, materials to win the war were available, but how many ranchers and farmers cared to use them? One wonders if all this effort was really necessary.

Harvester ants sting. Why do they sting? To defend themselves and, particularly, to defend the queen and nestmates. Harvester ants seem to have more different kinds of predators than other ant species and a glimpse into their lifestyle reveals why. The more populous and successful harvester ants live in open bare areas where they are conspicuously visible and where other prey are scarce. Their huge colonies with numerous individuals provide a bounty to any predator able to exploit them. An added disadvantage is that harvester ants live in the same colonies in the same places year in and year out and cannot easily move the colony to a safer location. Finally, harvester ants are relatively large, providing nutritional potential worthy of the attention of both larger and smaller predators. The sting is by no means the only harvester ant defense, as it is augmented by powerful biting mandibles, a variety of behavioral defenses, and alarm/recruitment pheromones.

In response to their disparate cadre of predators, harvester ants evolved individualized suites of defenses for each type of predator, often for each situation. What works against one type of predator frequently is woefully inadequate, or outright ineffective, against another type of predator. Stings work really well against you or me, but they are worthless against web-spinning spiders. Biting works well against other attacking ants but not against befeathered birds. That said, some generalities do emerge: Stings are generally useful against vertebrate predators and are not useful against insect and other arthropod predators. Biting mandibles are generally useful against insect and other arthropod predators or competitors but are of little use against large vertebrate predators. As is the biological rule, exceptions occur. In two instances, I have found dead vinegaroons (whiptail scorpions) killed by a harvester ant sting successfully placed in the intersegmental membranes of the vinegaroon's crushing, claw-like pedipalps (appendages near the mouth). Biting, a defense of little effect against many large vertebrates, is effective against horned lizard predators.

Social insects are superorganisms in which the entire colony acts much as a single organism, albeit one with separate mobile parts. As cells and tissues in the human body act for the good of the whole, social insect individuals act for the benefit of the colony. As our skin cells operate to protect the rest of our body, ant workers operate to protect the queen and the rest of the colony. The harvester ant superorganism is generally little affected by small insect and other predators, just as we are not seriously damaged by bedbug bites. Bedbug bites are a nuisance, for sure, and do kill some blood cells, but otherwise, they have little effect on our survival or productivity. Likewise, the loss of a few individual harvester ant foragers to small predators barely affects the overall survival or success of a harvester ant colony. Invertebrate predators threaten individual units, the workers, rather than the whole colony. A common invertebrate predator that takes an occasional worker is the ant lion, famous for making conical pits in sand and lurking out of sight in the sand at the bottom. If a hapless harvester ant slips down the loose sandy side of the pit, a pair of ice-tong mandibles await to skewer it. The ant appears to instinctively recognize the risk and rapidly attempts to run upward out of the pit. The ant lion responds with a shower of sand flipped on and above the ant, causing continual landslides of sand plus ant cascading back toward the awaiting jaws. Fortunately, most harvester ant individuals are too big and too fast, allowing them to escape the sand-slide torrent and flee from the pit. The ant lion will need to find a smaller ant next time.

A cadre of miscellaneous arthropod predators, including robber flies, assassin bugs, and a variety of spiders, ambush or trap a few harvester ant workers. Of this group, only a few spiders are of major potential consequence. Black widows and false black widows (*Steatoda*) can be most troublesome. False black widows have the temerity of building their webs directly over colony entrances of harvester ants and lurking out of reach a few centimeters above, ready to snag foragers. The success of this spider lifestyle rests on the extraordinary strength and stickiness of the silken web and on the ability of the spider to reside in its web safely above the fray. Although a spider will take only half a dozen ants per day—hardly enough to depopulate the colony—the ants respond strongly and negatively by moving their nest entrance or by shutting down all outside activity for days, thereby starving the spider and forcing it to leave. This heroic action by the ants seems economically counterproductive compared to lost foraging activity, although, perhaps, overall it is beneficial in ways not so obvious to simple economic bean counting. Web spiders, along with other predators inclined to reside near nest entrances, are believed to be a major reason western harvester ants clear all vegetation from around their colonies. Clearings remove anchors for the webs of false black widows and other spiders and expose these spiders or other predators to their own predators. These ant predators are less likely to risk their lives for just a few ants.[19,20,21]

Nature never ceases to deliver surprising inventions. One amazing surprise is hidden in the story of the digger wasp, *Clypeadon*, and harvester ants. Some species in the small genus *Clypeadon* prey exclusively on worker harvester ants. The wasp grabs a worker ant either near the nest or, when few ants are outside, enters the nest and attacks an individual ant inside and underground in its own lair. In either case, the ant is stung into a deep, permanent paralysis and carried back to the wasp's burrow. Despite having formidable mandibles that could easily dismember the wasp, the ants mysteriously rarely mount a strong defense against attack. We don't know why. Ants may not attack because they lack the ability to smell meaningful foreign odor from the wasp. Much as we cannot see a black object in the dark, the ants would not be able to smell an apparently odorless or tasteless wasp, and a threat unseen or not smelled is a threat undetected. This scenario seems impossibly strange to visually based organisms such as us, but vision plays no meaningful role in recognition by most ants. They rely on body "odor" signatures.

Clypeadon paralyzes and carries off between 16 and 26 ants to stock each larval cell of her nest. She lays an egg on one of the paralyzed ants, seals the cell, the egg hatches into a larva that eats all the ants, pupates, and then emerges as an adult. From this initial glance, *Clypeadon*'s life history description appears not that different from many other digger wasps. The departure from the usual digger wasp biology unfolds in the mechanism of transport of paralyzed ants. The standard digger wasp's method of prey transport is to hold the prey with its mandibles, its middle pair of legs, or stuck on its barbed stinger. Not *Clypeadon*. She flies off with mandibles, all six legs, and the stinger free and with the ant appearing glued to the tip of her abdomen. She has a unique structure, not possessed by males or any other wasps, appropriately called the "ant clamp." The wasp's clamp is composed of an enlarged specialized biconcave upper part of the abdomen matched to a mobile lower bilobed abdominal structure that clamps and locks against the ant's leg bases. This mechanism is ideal for unencumbered transport. Unfortunately, it is also ideal for lightning-fast parasitic flies to lay tiny maggots on the ant held behind the wasp. The maggots then appropriate the ants in the cell to make flies instead of wasps.[22] No new invention by nature goes unnoticed.

Remarkably few birds, and no mammals, have successfully exploited harvester ant colonies. George F. Knowlton of Utah State University devoted much of his life to studying vertebrate predators of insects, including harvester ants. Birds that at least occasionally might feed on harvester ants included rock wrens, sage thrashers, western meadowlarks, Brewer's blackbirds, sage sparrows, sage grouse, and red-shafted flickers.[23] How these birds avoided stings and bites is unclear—speed, agility, and, especially, slick feathers and hard crushing bills? Sage grouse, now an endangered species, eat the occasional western harvester ant and then add the insult of using their ant mounds as conspicuous high points to attract females to male strutting grounds.[24] Flickers display an interesting behavior with western harvester ants. They visit the conical ant nests in the morning when ant larvae and pupae are brought to the warmth of the sunny side and pick off the thin protective soil crust exposing the white larvae and pupae. The flicker is primarily interested in these white morsels, rich in protein and fat and low in fiber, though a few low-protein, low-fat, high-fiber workers get consumed in the process.[25]

Lizards are the most serious predators of harvester ants. Side-blotched lizards, long-nosed leopard lizards, and fringe-toed lizards eat worker ants, including harvester ants. Sagebrush lizards (*Sceloporus graciosus*) take the culinary fondness for ants a step further. Most analyzed individuals have harvester ants in their stomachs. Horned lizards, genus *Phrynosoma*, take eating ants to the extreme. Eighty-nine percent of the regal horned lizard's diet is harvester ants.[26] Horned lizards, squat, rotund creatures sporting a collar of vicious spikes at the back of their head, have long been favorite animals of people. Their rounded body shape and pathetically slow running ability make them easy to catch, properties that have allowed pet-loving people to love them to extinction in some areas. Among their unusual properties is an enormous stomach able to hold 13.4 percent of their weight.[26] No other lizard can match that feat. That's like a 200-pound person eating a 27-pound meal. As Japanese sumo wrestlers cannot sprint or run marathons, horned lizards lost their ability to run fast or far. Such is the price of becoming a harvester ant specialist.

If one is fat, slow, and eats mainly ants in exposed areas, some tricks are needed. An immediate trick that is not obvious until a horned lizard runs is its mastery of camouflage. They have color patterns closely matching the ground surface of their environment, and because they are short and broad, they leave no shadow. Some species even add fringes of scales to disrupt their body outline, thereby blending nearly perfectly with the substrate. Add to this very slow motion and enhanced awareness, and the lizard is rarely detected by either predator or ant. With a dietary necessity of up to a hundred ants per day, some tactical strategy is in order. The horned lizard strategy is to forgo frontal attack at the colony entrance, preferring to reside at the edge of a trunk trail or periphery of a colony clearing where individual ants can be picked off singly. A flick of the tongue and the ant disappears.

The fact that harvester ant venom is the most mammalian-toxic insect venom known led several of us to wonder how horned lizards could eat harvester ants with impunity. The venom of one harvester ant is sufficiently potent to kill a horned-lizard-sized mouse several times over. How, then, does the lizard avoid a lethal encounter when eating a hundred ants? Wade Sherbrooke, the foremost horned lizard biologist, my wife, and I put our heads together (probably over some beers) to determine whether we could answer this question. Wade was completing his doctoral dissertation on the mechanism in horned lizards of skin color change to match the background. I asked him his method.

"I just sacrifice a lizard and incubate its skin for hormone and other analyses."

"What do you do with the rest of the lizard?"

"Oh, I just discard it."

"Aagh! You throw it away and waste all the rest of the lizard? You can't do that. Give me the blood."

And so began the project. Wade provided lizard blood, my wife, a master of dissecting harvester ants for their venom, dissected thousands of ants, and I focused on determining the physiological mechanism of lizard tolerance to ant stings. The first question was, Are horned lizards susceptible to harvester ant venom? When tested with enough venom to kill 100 mice, the lizard was totally unaffected. That means they are not susceptible to the ant's venom. So, get out the shovels and buckets, we have work to do. Back from the blazing Arizona sun with several more buckets filled with Maricopa harvester ants, we began the task of harvesting more, much more, venom. That done, we finally found a median lethal dose for horned lizards: more than 1,500 times the lethal dose for a mouse was required to kill a horned lizard, an amount that translates into the venom from 200 ants. Jarrow's spiny lizard, a relative of horned lizards, was much more susceptible to the venom than horned lizards. Newborn horned lizards living in an area lacking harvester ants were about as resistant as lizards feasting on the ants. These findings indicated that horned lizard resistance was innate and not induced immunologically. Horned lizards were not undergoing self-vaccination through diet; they were eating harvester ants with impunity because they have a blood factor that neutralizes the ant venom. This was confirmed by classic antivenom + venom studies in which lizard blood plasma was mixed with 3.6 times a lethal dose of venom and injected into mice. The mice showed no adverse reactions. Horned lizards are the first known example of a vertebrate predator that has evolved an innate resistance to an insect venom.[27]

The harvester ant–horned lizard story doesn't end here. Both the horned lizard and the harvester ant have more surprises. Horned lizards secrete a special slippery and viscous mucus that lines their mouth and digestive system. When an ant is eaten, its sting, rather than piercing the delicate throat or stomach lining, usually slips off harmlessly. The ants, for their part, have two defenses. Ironically, although the lizard isn't bothered by stings, it is intolerant of bites. If the ants become alerted to the lizard, they release alarm pheromone from glands at the base of their mandibles to recruit other ants to initiate a mass attack. This mobbing attack drives off the lizard, while providing an added bonus of exposing the fleeing lizard to the risk of being spotted by a roadrunner or another of its own predators. As a final insult to the lizard, the ants can bite and clamp so permanently to the lizard's toes or soft underbelly that the head remains attached long after the ant has died, as a reminder of the ant's ability.

One species of harvester ant, the enigmatic *Pogonomyrmex anzensis,* lives in such a harsh environment that horned lizards do not live there. This ant, described from a single collection in the Anza-Borrego Desert of California, was subsequently lost to science for 45 years despite an intense search by some of the finest myrmecologists of the time. The "mystery" ant was rediscovered in 1997 by Gordon Snelling after years of careful search. The ants nest in an unimaginably harsh location among south-facing, hard, rocky hillside slopes that are exceptionally sunbaked, hot, and dry. No horned lizards live there; in fact, Gordon could find no potential predators living there. We asked, What happens to the venom of a species that has no vertebrate predators? To answer this question, Gordon collected these ants and shipped them to me where I ran some comparative tests. The venom sacs of these ants were collapsed and mostly empty, containing only about one-sixth the venom expected. The sacs were of normal size. They just contained little venom. Gordon also noted that the ants were very timid and could not be forced to sting readily. When analyzed for toxicity, the ants surprised us. They were among the most toxic of any harvester ants, some three times more toxic than Wheeler's harvester ant, the largest and one of the most aggressive of all North American harvester ants. Apparently, the evolutionary response in *P. anzensis* to a harsh environment and the loss of major predators is not to lose or sacrifice venom activity but, rather, to spare the energy required to make venom by not producing much venom.[28]

Lizards are not the only animals that eat hundreds of harvester ants at a sitting. Historically, people have also. One fine afternoon in 1994 I received a call from Kevin Groark, a master's student at the University of California, Los Angeles, who was studying cultural traditions of California native peoples before the massive influx of the Anglo population. The young men of several tribes engaged in spiritual trips called "vision quests" in which they sought "dream helpers" to enhance their lives or to provide strength. Before they embarked on a vision quest, they fasted and vomited, often for several days in preparation. Vision quests were generally conducted in winter, which added another stress. When ready, they would ingest harvester ants in balls of eagle down feathers delivered by elder women. Kevin wondered whether the toxic nature of the venom could "poison" a youth into a state of hallucinations during which the religious spirits entered his body to give him guidance or strength.

"Heavens no," I replied. "A toxic dose of harvester ant venom for a person would be near a thousand stings. Perhaps the extreme pain alone could send them into a trance."

Kevin replied that they eat a lot of ants.

"How many?"

"Three hundred fifty or so."

"Wow. *That* could cause a sublethal effect that, combined with the rest of the ritual, might well induce hallucinations."

I can't imagine the pain, much less the sheer determination, required to go through such a ritual. On the basis of careful field notes by early anthropologists and voucher specimens, the ants in question were identified as California harvester ants.[4] As if the vision quests were not painful enough, some Indian groups "hardened" youths through puberty rites, as described by John Harrington in 1933, in which the early adolescent was "whipped with nettles and covered with ants that they [the boys] might become robust. This infliction was always performed in summer during the months of July and August when the nettle was in its most fiery state. . . . [The boy was whipped] until he was unable to walk. He was then carried to the nest of the nearest and most furious species of ants, and laid down among them, while some of his friends, with sticks, kept annoying the insects to make them more violent. What torments did they not undergo! What pain! What hellish inflictions! Yet their faith gave them power to endure all without a murmur, and they remained as if dead."[4] Perhaps fortunately for the ants, young men in California no longer have such rites.

Harvester ants are clearly special no matter from what angle they are viewed. Their stings and venom are particularly unique and extraordinary. Exactly what made them special captured my interest from my first experience. Love at first sting! Harvester ant stings were well noted and documented in the ant literature.

H. C. McCook wrote in 1879 that harvester ants are "regarded with a wholesome fear by children, and adults have little wish to meddle with them. . . . I tried to engage him [a bright, stout youth] to assist in the digging, but my offer was rejected with an emphasis and facial expression of horror which was amusing. . . . 'I wouldn't do dat fur five dollas a day [= $845 in labor costs in 2014]'!"[15] He continues that after the sting

> there was a sharp, severe pain, resembling that of the sting of a bee. Then followed twice, at short intervals, a nervous, chilling sensation, which seemed to sweep upwards, and was felt quite sensibly around the roots of the hair. This was a very peculiar feeling indeed, and appeared to me very like that caused by a sudden alarm, or excitement in which the element of horror predominates. Then followed a steady heavy pain about the wound, which continued for three hours more or less severely, a slight numbness accompanying. . . . Her sting pained me very much indeed, and was felt more than twenty-four hours afterwards.[15]

D. L. Wray in 1938 described his sting reaction to a Florida harvester ant this way: "It turned deep red in color and immediately a watery, sticky secretion came out of the skin. It beaded out like heavy perspiration and ran down my arm. This area became hot and feverish and the excruciating pain lasted all day and up into the night."[29] Creighton in his classic 1950 "Ants of North America" writes matter-of-factly "the sting of most species of *Pogonomyrmex* is exceedingly painful. It is not a localized reaction, like that of a bee sting, but one which spreads along the lymph channels and often causes intense discomfort in the lymph glands of the axil or groin long after the original pain of the sting has ceased."[1] Arthur Cole in the introduction to his 1968 *Pogonomyrmex Harvester Ants* detailed a sting: "The effect of a sting can be very painful. Localized swelling and inflammation ensue rapidly. Soon thereafter a throbbing pain, which may last several hours, extends to the lymph nodes of the inguinal, axillary, or cervical area, depending on the location of the sting. Frequently, the skin around the wound becomes very moist." Finally, the great husband and wife team of George and Jeanette Wheeler wrote in 1973:

> [The author was] stung on middle of free edge of upper lip. The ant was knocked off promptly . . . [the] pain did not begin for ten minutes—a general burning sensation. After an hour a dull ache commenced in the lips, the incisor teeth and the adjacent jaw. The middle third of the lip was slightly swollen and slightly red. . . . Six hours after the sting the ache had subsided to be followed by a burning sensation in the upper lip. The next morning (ten hours after the sting) there was a diminished burning sensation in the middle 55 mm of the free edge of the lip, which was edematous but not reddened; the middle 10 mm of the 55 mm was insensitive to touch. After 12 hours feeling began to return but the free edge was still numb, slightly swollen and with a slight burning sensation. After 24 hours the swelling was scarcely visible but the lip felt tight. There was no pain. . . . Two days after the sting the inner surface was hypersensitive and felt

> hot when rubbed. . . . Twenty-six days after [the] sting the middle section of the lip was still hypersensitive to touch.[30]

From these descriptions, we see that nobody who is ever stung by a harvester ant is unaffected, and the consensus is that these stings are remarkable in more ways than we wish to repeat.

Harvester ant stings differ from those of all other known insect stings in at least five ways. First, and to the victim's disadvantage, is the lack of an instantaneous reaction to the sting. No lightning bolt of pain, no fiery ember burning the skin. Rather, a somewhat delayed reaction before the pain is noticed and then the inexorable increase in pain. How long this initial "painless" period lasts after insertion of the sting is hard to know because if we do not realize we are being stung, we cannot start the stopwatch. When we do notice, it is too late; the damage is already done. My feeling is that the unnoticed period, at least for stings to the feet or legs, is about 30 seconds. How does this delay in pain benefit the harvester ant? After all, isn't it to the ant's benefit for the pain to be immediate? The answer might reside in short-term versus long-term benefit for the ant and her colony. The ant's venom injection system is rather slow. Thus, if instantaneous pain were produced at the very beginning of the process, the victim could remove the ant and quickly stop the damage. However, if a delay in pain occurs, then the ant can deliver a richer, fuller dose of venom to maximize long-term damage.

A second difference between harvester ant stings and other stings is the localized sweating induced around the sting site. The sweating seems more viscous and sticky than usual sweat and does not occur anywhere else on the skin. This sweating is easily detected by gently moving a finger from side to side across the sting area. The finger glides smoothly and easily over the nonsweaty area, then encounters a friction in the sweaty area, and finally returns to gliding smoothly and easily again. To make the test more sensitive, I sometimes use my upper lip instead of a finger. No other insect sting causes sweating around the sting site. Therefore, should the ant escape unnoticed, sting site sweating is diagnostic for a harvester ant sting.

A third difference between harvester ant stings and other stings is localized piloerection, the standing up of the hairs around the sting site. The hairs immediately around the sting puncture stand up much like the hairs on the shoulders of a frightened dog. In addition, "goose flesh" in the area caused by the contraction of single-celled erector muscles at the base of each hair produce a dimpling appearance of the skin. As with sweating, the hairs outside the sting area are of normal appearance. No other insect sting causes localized hairs to stand up; this feature is, again, diagnostic for a harvester ant sting.

A fourth difference between harvester ant stings and those of other insects is the generation of lymph node pain. The nearest lymph nodes—the axial in the armpit region for stings to arms or the inguinal in the groin area for stings to the legs—become tender and hard. The pain is not sharp or unbearable but is decidedly and noticeably unpleasant. The feeling is hard to describe. I call it "unaesthetic" because it interferes with one's sense of happiness and well-being. No other insect sting causes lymph node pain or awareness and is diagnostic for a harvester ant sting.

The fifth difference between harvester ant stings and those of all but one other stinging insect is the length and nature of the pain. Once the pain kicks into high gear in about 5 minutes or so, it continues unabated for hours. It comes in a wave with intense, excruciating, teeth-gritting pain and then recedes to a more manageable level before returning to a new peak, and so on. Unfortunately, the pain does not go away quickly. For most people, it lasts 4 to 12 hours, depending on the species of harvester ant, the amount of venom the ant managed to inject, and the sensitivity of the individual to pain. The pain of rough, red, or western harvester ant stings lasts about 4 hours, whereas the pain of Maricopa, California, or Florida harvester ant stings lasts closer to 8 hours. And I chose to focus my research on Maricopa and Florida harvester ants!

From the unusual nature of harvester ant sting reactions, I suspected the chemical composition of harvester ant venom must be different from other insect venoms. At the point in the 1970s of my first stings, knowledge of harvester ant venom chemistry was a blank page. The first insect venom chemically characterized was from the horse ant, or English red wood ant, *Formica rufa*. In 1670, 90 years before Linnaeus gave the ant its scientific name, John Wray determined these ants contained formic acid.[31] The acid lacked a name and, subsequent to its discovery in ants, was called "formic acid" after the Latin *formica*, meaning ant. Wray's discovery was quite the topic of discussion in scientific circles. Even the public was aware of it. Thereafter, the public and many scientists came to believe that (all) insect venoms were formic acid, a belief that sticks today. Urban myths, such as this formic acid myth, are nearly impossible to dispel. German scientists, starting in the late nineteenth century and into the 1930s, essentially showed that honey bee venom contained highly active solid toxins, unlike the liquid volatile formic acid, and never mentioned an odor of formic acid, something that would be conspicuously obvious if it were present. Likewise, no venomous ants that have functional stingers (formic acid–producing ants lack functional stingers and can only bite) have ever released a formic acid odor when collected or aspirated into a jar. The myth of formic acid in insect venoms is likely to remain part of our street knowledge and to entertain us for decades.

Because nothing beyond descriptions of sting pain was known about harvester ant venom, my first task was to collect venom for chemical and pharmacological analysis. The ants themselves are easy to collect, but they have only a tiny amount of venom. Each ant produces about 25 micrograms of venom. For an ounce of venom, over 1 million ants would be required, and at 3 minutes per ant, that would require six and a half years of working nonstop, day and night, to collect. Obviously, collecting large quantities of venom was out of the picture. Fortunately, many analyses require only tiny amounts of venom. Enzymes that cleave chemical bonds, thereby creating biochemical havoc within organisms and their body tissues, are one of the effective ways venoms operate. Florida harvester ant venom is a cornucopia of enzymes with more different highly active enzymes than any other insect venom. These enzymes do various things once injected through the stinger into the skin of something like my foot. Two phospholipases, A1 and B, break phospholipids in cell membranes, in the process releasing the fatty, pain-inducing lipid lysolecithin and other fragments, and destroying the cell membrane. Ouch. Another enzyme, hyaluronidase, acts as meat tenderizer to soften the connective tissue in the skin, thereby facilitating entry of other venom components so they can wreak their havoc. Other enzymes, including esterase and acid phosphatase, break down other molecules in the skin and body, thereby synergizing the activity of other venom components. Whether these are outright toxic on their own is unknown. One last enzyme is particularly intriguing. It is lipase, an enzyme that breaks bonds in fat molecules.[32,33] The exact function of lipase, an enzyme unknown from any other venom, remains a mystery, but my suspicion is that it, perhaps with esterase, causes a sharp, rashy feeling like that produced by stinging nettle plants.

Active direct pharmacological and toxic actions are another hallmark of harvester ant venom.[34] The venom is highly hemolytic, rapidly destroying the membranes of red blood cells. Destroyed blood cells release their hemoglobin with the combined effect of impairing oxygen transport in the body and clogging the kidneys' filtering system. Without functional kidneys, one dies a painful death over several days. Hemolysis can be an indirect cause of death.

Kinins are highly active peptides most famous for affecting cardiac activity, lowering blood pressure, and causing pain. Wasp kinins appear to be the main cause of the pain from yellowjacket and other social wasp stings. Kinin-like activity is pharmacologically detected in harvester ant venom, although the exact chemical structure of the molecule that produces the kinin-like activity has not been determined.[35] The effect of this activity in a sting reaction is unclear, likely causing short-term pain.

Harvester ant venom's most profound activity is its direct neurotoxicity. This neurotoxicity directly targets nerves in the skin, spinal cord, brain, and, likely, the heart. Near instant death can be the result. Small vertebrate predators are at serious risk from even a few harvester ant stings. Fortunately, for people, the amount of venom in a few stings is too low to have a meaningful toxic effect. In addition, our thicker skin is a marvelous slow-release system for venom. The venom from a few stings gets stuck in the skin and is only released at rates below that which would do systemic damage.

Venom lethal toxicity is measured as the median venom dose required to kill half of the victims. In contrast to snake venoms where lethality values were well known, the only insect venom for which we had a value was the honey bee. Honey bee venom is remarkably toxic, surpassing the toxicity of many snake venoms. To our surprise, when harvester ant venoms were analyzed, their potencies dwarfed that of honey bees. The venom of an average species of harvester ant is 6 times more deadly than honey bee venom, and the Maricopa harvester ant from Willcox, Arizona, is about 20 times as toxic.[36] To date, harvester ant venom is the most toxic known insect venom, more toxic, even, than all snakes, except for a handful of Australian and sea snakes. If harvester ants were the same size as sea snakes, I suspect we would know a lot more about their venom.

In their quest to defend the colony, some harvester ant species have another trick in their magician's toolbox. A major limitation of many insect sting systems is the speed of venom delivery. Ants and bees lack forceful mechanisms for near instantaneous venom delivery. A major benefit of rapid venom delivery is getting a large quantity of venom into an assailant before being discovered and removed or killed. Harvester ants in the California species group and in the Florida ant have a marvelous system to get around this venom-delivery problem. They simply autotomize their stings into the flesh of their target. In so doing, the ant can be removed, squashed, eaten, or what have you, yet the autotomized and intact venom system remains and continues unobtrusively delivering its full payload of venom. Sting autotomy is familiar in honey bees and is considered diagnostic for bee stings: If a stinger is left in the skin, it is a honey bee. If a stinger were not present, then it is something else. This wisdom is only partially true; harvester ants and several, mostly tropical, species of wasps also leave their stings embedded in the flesh. The individual ant dies, but its full venom delivery helps protect the colony superorganism, including queen, immatures, and other adults. Because worker harvester ants are reproductively sterile, the suicidal ant sacrifices no direct reproduction, and her strategy of total venom delivery maximizes the chances that her close kin, the queen, brothers, reproductive sisters, and colony they live in, have an enhanced probability of reproducing.

This brings us to the sting of harvester ants. Harvester ant stings are definitely not for the faint of heart, making the beekeeper's task appear like a stroll through the meadow. Harvester ant stings are serious, bad, and painful. They might start out innocently at first, perhaps feeling a bit like someone is injecting tiny amounts of water through a dentist's syringe into the tender spot, but the sensation soon morphs into a sharp, digging pain. The pain is sometimes like the dull, heavy thud of being hit with a lead-filled blackjack; other times like a wizard is reaching deep below the skin and ripping muscles, tendons, and nerves. Except the muscles, tendons, and nerves are not ripped only once but in waves: ripping now, easing a bit, ripping again. To make the point even clearer, the torture continues unabated for several hours. Expect 4 to 8 hours of pain. On the pain scale, harvester ant stings command the very high level of 3, substantially higher and worse than that of a honey bee sting. These stings beg for stories to loved ones and friends as gestures of care and compassion.

9

TARANTULA HAWKS AND SOLITARY WASPS

I would rather be stung a hundred times by digger wasps than once by that darling of the philosophers, the honeybee!
—Howard Evans, *Wasp Farm*, 1973

Electric. A bolt out of the heavens. That is what tarantula hawk stings feel like. The question is not are tarantula hawk stings different from those of other stinging insects, but why and how they are different. "Why" questions present a problem for science, as they suggest some purposeful reason behind the observation, and purposes are not amenable to scientific methods. For the moment, if we ignore that limitation and view the "why" as a catalyst for liberating our minds to generate ideas that explain our observations, then we open the door to understanding. Ideas that flow through the door can then be tested and, with good fortune, lead to understanding.

A good start to searching for the whys of tarantula hawk sting pain is in the biology of tarantula hawks and other "solitary" wasps. Solitary in the sense of wasp biology means lack of sociality, that is, not living in colonies with sisters, brothers, mothers, and growing young. Instead, solitary wasps live a life of the single female who must do everything herself so that her offspring survive and carry on her lineage. Solitary wasps are true single moms. Male wasps do no work whatsoever to assist in producing the next generation. Males only mate with females to provide the necessary sperm; otherwise, they are mostly a nuisance to females in their continuing, amorous courting. In some bees, males actually provide some benefit by guarding the nest entrance and by blocking it with their head to prevent intrusion by parasites or other usurping female bees. These males truly are using their heads. Unlike bees, no solitary wasps share a common nest with other females, and males do not use their heads to protect the fort.

Solitary wasps are generally predators that actively hunt and overpower prey. Bees, however, are vegetarians, sipping nectar or other sweet liquids and munching on pollen from flowers. One peculiar group of bees, the *Scaptotrigona,* commonly called "vulture bees," abandoned pollen collecting in preference to scavenging meat from dead animals. These bees are the exception to the vegetarian bee rule. Solitary wasps have their own exceptions to their predation rule. The exceptions are the pollen wasps in the subfamily Masarinae, relatives of the potter wasps, named for their often exquisite pottery vases in which the young are reared. Pollen wasps, as the name suggests, collect pollen to provide for their young instead of prey. They are an example of convergent evolution with bees, as they are not closely related to bees, residing on a different branch of the evolutionary tree of life.

If you are a wasp, solitariness has several disadvantages compared to sociality. No, we are not referring to social life in the sense of human social life—friends, parties, shopping together, shared meals, mutual laughs. We are referring to lacking the advantages of siblings and the community effort to enhance life's prospects. Social insects have multiple individuals to work on a task and can usually do the job better than any one individual. For example, one member might find a food bonanza, such as a large, dead grasshopper or a new patch of flowers. The individual can only harvest a small portion of the resource alone and might lose all of it to competitors. A social individual can recruit others to dominate and harvest the large resource. A member of a solitary species that fails to find food or water can be in dire straits. If a member of a social species fails in foraging, usually others succeed and can share the take with the individual that failed. Solitary individuals must be jills-of-all-trades and do everything themselves. In social species, individuals can specialize: some collect food, others collect water, some collect resources for nest building, others feed and care for the young, and still others defend the nest. This partial list is only an illustration of potential specializations and is far from complete.[1] One important advantage of sociality is the ability to have someone always at the nest to defend it from predators, parasites, or intruders. Solitary wasps must leave the nest to secure food, prey, and water, leaving the house open with nobody to guard the baby.

Solitary life for wasps is a trade-off between the advantages of sociality versus the advantages of solitariness. One advantage of solitariness is timing. If food is present in abundance only during a brief time, for example, when young katydids are present, then a solitary wasp species needs to be present only during that time. The wasp has no need to be active or present during the rest of the year. This saves energy and reduces risk from predators and the abiotic factors of weather and climate. Solitary species also can become super niche specialists that do one thing better than anyone else. For example, the colorful striped wasp, *Cerceris fumipennis*, specializes in finding, capturing, and paralyzing rock-hard metallic wood borer beetles and in finding them efficiently.[2] In contrast, entomological experts studying these same beetles have great difficulty locating them and, in several cases, have turned to mining nests of *Cerceris* to find beetle species new to science. Another example is the cicada killer wasp, which readily finds and captures cicadas, something social wasps and people cannot frequently achieve.

As parents know well, preschool children are prone to catching and transmitting contagious diseases. Contagious diseases of many varieties also plague social insects. Solitary species, thanks to their low contact with one another, avoid many problems of infectious diseases. Another major advantage of the solitary lifestyle is the ability to be less obvious to predators. The bustling activity of a colony and its often large nest are more apparent to hungry predators than a solitary wasp nest. Solitary wasps with their infrequent activity are far less likely to be noticed or to draw unwanted attention. Importantly, an individual solitary wasp is simply not worth the effort for many large predators to capture and eat or to dig up and destroy its nest to consume provisions and young. A large predator is more likely to view numerous individuals in a nest as worth the potential pain and effort. A human equivalent is the enticement of a person to cross a room for a single peanut versus the lure of a bowl of peanuts.

An associated defensive option more available to a solitary species than social species is the advantage of fleeing from predators. When workers of social species are near their nest, they tend to defend and hold the fort against predators. The option to flee is reduced for the worker. If the worker doesn't prevent the destruction of the queen(s), all, or most, is lost. If the nest of a solitary wasp is attacked, the wasp simply flees, saving herself. Even if the nest is destroyed, an unlikely event, considering most predators might not recognize the nest presence or decide it is of too low value, the wasp can readily build another and continue.

The greatest risk of life is living. The expressions "live and let live" and "eat, but don't be eaten" capture the essence of the problem. Life is all about the family, the individual's family, and ability to continue the family into the future. Life is required for reproducing the family into the future, but the process of living is a major risk. At any time, life can terminate by ending up in someone's stomach. The longer one lives, the greater the risk of being eaten. One solution to life and family is to reproduce quickly and then, the job over, to die. Mayflies exemplify this extreme. Some live as adults less than an hour, mate, and die on the water's surface with eggs pouring out as they die.[3] Nature selects for maximal efficiency of reproduction with the minimum time devoted to activities that do not directly enhance the reproductive effort. For many solitary wasps, short life is an adaptation to avoid risky living and non-reproductive time exposure to predators. If the wasp is a prey specialist, as most are, and the prey is present each year for only a few weeks or so, then the best life strategy is to emerge as an adult wasp just as the prey presence is beginning, to work feverishly to catch prey while it is available, to reproduce, and to die after the prey season ends.

No point in staying around beyond that. Many social species do not have the option of short adult life span. They must maintain the colony throughout the year, including long seasons of total, or near-total, unproductivity, all the while exposed to predators.

Stung by a tarantula hawk? The advice I give in speaking engagements is to lie down and scream. The pain is so debilitating and excruciating that the victim is at risk of further injury by tripping in a hole or over an object in the path and then falling onto a cactus or into a barbed-wire fence. Such is the sting pain that almost nobody can maintain normal coordination or cognitive control to prevent accidental injury. Screaming is satisfying and helps reduce attention to the pain of the sting. Few, if any, people would be stung willingly by a tarantula hawk. I know of no examples of such bravery in the name of knowledge, for the reputation of spider wasps—specifically tarantula hawks—is well known within the biological community. All stings experienced occurred during a collector's enthusiasm in obtaining specimens and typically resulted in the stung person uttering an expletive, tossing the insect net into the air, and screaming. The pain is instantaneous, electrifying, excruciating, and totally debilitating.

Howard Evans, the great naturalist and author of the eminently enjoyable books *Life on a Little Known Planet, Wasp Farm,* and *The Pleasures of Entomology,* was an expert on solitary wasps. Howard, a slight, reserved man with a shock of white hair and a sparkle in his eyes, was especially fond of tarantula hawks. Once, in his dedication to the investigation of these wasps, Howard netted perhaps 10 female tarantula hawks from a flower. He enthusiastically reached into the insect net to retrieve them and, undeterred after the first sting, continued, receiving several more stings, until the pain was so great he lost all of them and crawled into a ditch and just sobbed. Later, he remarked that he was too greedy.[4]

I know of only two people who were "voluntarily" stung by tarantula hawks. I say "voluntarily" as both were film actors performing their duties, which, among other things, "encouraged" being stung. One was a young, handsome athletic entomologist who knew of the wasps. He deftly reached into the large cylindrical battery jar and grabbed a wasp by the wings. He had her in such a position that her sting harmlessly slid off his thumbnail. We prattled for a minute or so about tarantula hawks while the camera scanned close up to the long sting as it slid harmlessly, missing its mark. Then with a great heave the wasp pulled its abdomen back and thrust the sting under the nail. Yeee...ow (we can't recall if any expressions unsuitable for general audiences were uttered), the wasp was hurled into the air and flew off unharmed. One point for wasp, zero for human.

The other actor was a solidly built fellow who could easily have been a football linebacker, and who was a master of performing pain-defying acts of bravery. I, however, was charged with catching the wasp and delivering it to the scene. Five or six tarantula hawks were easily netted from flowers of an acacia tree; unfortunately, the net snagged on some thorns, and all but one wasp escaped. The remaining wasp was a male, so I summoned the cameraman to demonstrate how males cannot sting and are harmless. I reached in and casually grabbed "him." The him was a her. Yeee...ow, except this time it was me. I managed to toss her back in the net, while attempting to explain my blunder and pain on camera. Unfortunately, I was not the actor, so the footage was relegated to some obscure studio archive, perhaps someday to be resurrected on YouTube. That episode over, the tarantula hawk was delivered to the rightful actor. He grabbed her, was stung, and showed no reaction beyond a begrudging "ouch, that did hurt a bit." I figured the guy had no nerves. But his director handed him a habanero pepper, the tarantula hawk of chili peppers, which he enthusiastically bit into. He became instantly speechless, convinced fire was blasting from his mouth, nose, and ears. Apparently, he did have some nerves—sensitive at least to chili peppers.

Tarantula hawks have never been recorded as a part of human warfare, but they might be candidates in some future altercations, and surely they come a close second in personal battles. Howard Evans, in a moment of exasperation, wrote of an experience in Mexico: "[Tarantula hawks] are spectacular creatures, on a number of occasions I collected these wasps in the Southwest and in Mexico, followed by a group of urchins who asked questions and tried to help. My trick to be rid of them was to pick a tarantula hawk off the flowers with my fingers and show it to them. Of course I always picked up a male, which cannot sting. But my curious followers would pick up a big one, usually female, and quickly decided they wanted no more of that."[5]

How could such a small animal as a tarantula hawk embed itself so strongly in the human psyche and win? Several years ago I attempted to address this question in a paper entitled "Venom and the Good Life in Tarantula Hawks: How to Eat, Not Be Eaten, and Live Long."[6] The natural history of tarantula hawks provides some insights. Tarantula hawks are the largest members of the spider wasp family Pompilidae, a family of some 5,000-species strong that prey solely on spiders.[7] The feature of tarantula hawks that makes them so special is their choice of the largest of all spiders, the fierce and intimidating tarantulas, as their target prey. The old saying "you are what you eat" rings true for tarantula hawks: if you eat the largest spiders, you become the largest spider wasps. As with other spider wasps, the female wasp provides each young with only one spider that serves as breakfast, lunch, and dinner for its entire growing life. The law of supply and demand applies: large spiders produce large wasps; small spiders produce small wasps. The story doesn't end here. Momma wasp is not entirely at the whim of fate and fortune in the size of spiders she encounters, with her young randomly enduring the consequences. She has the special ability to choose the sex of her babies. Hymenoptera are oddballs in the genetic world. Females are produced from fertilized eggs, and males are produced from unfertilized eggs. This not only means males have half the genetic information of females (but that does not translate into males being half as intelligent as females, a thought that might enter the human female mind), it also means mom can choose to produce a son or a daughter by selectively allowing stored sperm to fertilize the egg. In the tarantula hawk world, females are valuable. They do all the work, take all the risks of capturing the spiders, and have to drag a spider sometimes eight times their weight to their burrow. Thus, females need to be big and strong to do the job efficiently and to produce the most young. Males, however, mainly sip nectar from flowers, chase other males, and mate with females. A small male can mate with a female, so size is not so crucial, though a bigger male is usually more successful in winning more females. Mother tarantula hawks choose to give the valuable resources of large tarantulas to female young and small tarantulas to male young.

Tarantula hawk life history is similar to that of many other solitary wasps. Female adults emerge from their underground cells to seek nectar for food and to mate. Males emerge to seek flowers and to begin mating behavior. Male tarantula hawks in the genus *Hemipepsis* are famous for their hilltopping behavior. They go to hilltops, ridge lines, or other prominent high points and establish leks, or mating territories. In these territorial leks, males battle other males to defend their territories, with larger males usually winning the best territories, often near the center of the lek. Virgin females visit the lek to seek mates. They mate briefly once in their lifetime and get on with life. In Arizona, mating systems for tarantula hawks in the genus *Pepsis* appear to be centered around preferred flowers, especially some milkweeds, western soapberry trees, or mesquite trees. The males actively patrol these resources; otherwise, mating is likely equally ephemeral as in *Hemipepsis*. Mated females then embark on finding tarantulas. They tend not to be fussy and take tarantulas of several species, both male and female tarantulas and adults and large immatures. Large, plump, juicy female tarantulas are mostly destined to become food for baby female tarantula hawks. Male tarantulas are scrawny, long-legged creatures that usually weigh much less than females. They mostly are destined as food for the next-generation male tarantula hawks; hence male tarantula hawks are often tiny compared with their sisters.

Tarantula hawks sting their tarantula prey between a leg base and the sternum, the plate between all the legs. The sting, directed at the large nerve ganglion that controls the legs and fangs, completely inactivates and permanently paralyzes the spider within one-and-a-half to two-and-a-half seconds. The now limp spider is dragged to a nest burrow, constructed by the female wasp, or to the tarantula's own nest burrow. Anyone fortunate to witness at dusk the unfolding of one of nature's great dramas, as a tarantula hawk drags her enormous spider long distances over the ground, is treated to an adventure remembered for a lifetime. The spider is placed in a cell at the bottom of the nest tunnel, an egg is laid on the spider, and the tunnel is filled with dirt and sealed. The mother's duties now done, she is off for another prey. The egg hatches in a few days into a first instar larva that imbibes blood from the live, paralyzed spider. Over the next 20 to 25 days, the larva grows, molting its skin four times, and finally becomes a fifth instar larva. The spider is still alive at this point, even though the larva has eaten blood, muscle, fat, digestive system, and reproductive system, leaving the heart and nervous system. The fifth instar now rapidly consumes the rest of the spider before it can spoil. With the food exhausted, the larva spins a silken cocoon and pupates. In early season, the pupal stage may be only several weeks, after which the adult emerges. Later in the season, the wasp overwinters in the cocoon phase to emerge the next spring. Adult males live a few weeks or so, whereas females can live four to five months.[8]

One of the great mysteries is why tarantulas do not fight back when a tarantula hawk attacks. Why a huge spider whose massive, powerful fangs easily crush an enormous cockroach, usually with a satisfying crunch, or a hard beetle, fails to defend itself against a wasp is unfathomable to the human mind. How can the tarantula passively submit to its killer usually without the tiniest defense? We cannot journey into the mind of a spider to answer this question. This question also pertains to the spider prey of most, if not all, of the thousands of other spider wasp family members. Perhaps escape and freezing is a better defense in the long run than fighting. How the tarantula distinguishes a tarantula hawk from a cockroach or beetle is also not known, although some ideas seem reasonable. Unlike humans who "see" the world through our eyes and, secondarily, through our ears, with touch, taste, and smell minor modalities, spiders, insects, and most other invertebrate animals sense the world primarily through smell and secondarily with touch and some vision or hearing. In spiders and insects, smell includes contact receptors on antennae, pedipalps, legs, and other body parts. Many of these receptors detect surface chemicals on their prey. The surface chemical blends provide a signature of their source that the insect or spider can recognize. To us, a wasp smells pretty much the same as a beetle, a moth, or a fly—that is, it has no odor—but to a spider or an insect, they are distinct. The tarantula, a nearly blind animal, likely recognizes the tarantula hawk primarily by its odor, perhaps aided by the wasp's "feel" and the vibrations it sends through the soil surface or air pressure waves. The wasp might also be recognized by the distinct airborne odor it releases. This distinctive odor is easily detected by humans, especially when the wasp is captured or threatened. The odor is pungent but not harsh, acrid, or repulsive (as in a dead animal or overflowing sewer). The odor is distinctive and somehow seems to get into the human psyche, engendering a strong repellency.

Naturalists frequently commented on the odor. Alexander Petrunkevitch, an eminent early arachnologist at Yale University, noted one tarantula hawk, when contacting the jaws of a tarantula, "raised her wings and suddenly produced a rather pungent odor," which he concluded "the production of this odor must be a sign of anger, perhaps of warning."[9] F. X. Williams, the man who probably studied tarantula hawk behavior more than any other person, described the odor as "the *Pepsis* odor," a familiar and universal odor among *Pepsis* species.[8] Howard Evans noted that both males and females of *Pepsis* "have this characteristic odor, and this odor may well be repellant to predators."[5] Sadly, although we know the odor is produced in the mandibular glands, so named because they are at the base of the mandibles, we have not identified its chemical nature. This is not from lack of trying. I have worked with five or six excellent chemists over more than three decades, and we have failed to solve this mystery. Also mysterious is the role, or roles, of the odor. The most evident role is chemical defense against predators, including entomologists, that try to catch them. This defense is not direct, as with carpenter ants that spray corrosive formic acid on assailants or blister beetles (the group containing "Spanish fly") that cause painful, vesicating skin rashes; rather, it appears indirect in the form of an aposematic odorous warning to stay away, or else. The honesty of this warning defense is obvious to anyone who has grabbed a female tarantula hawk. The odor also might be an aggregation pheromone that attracts both males and females to rich floral sources, to resting places to congregate, or to lekking areas for mating. Finally, the odor might serve to flush a tarantula from its burrow and/or prevent the spider's natural defensive behavior.[10] As often occurs in biology, the odor might have first evolved for one role and later been selected for several additional roles.

Let's return to the question of why the tarantula does not fight back. Could it be that the wasp somehow inactivates the spider's defenses or frightens it with its motion, wing buzzing, or odor that the spider is paralyzed with fear? Such a concept surely seems too wild-eyed to be real, but who knows. We have little understanding of fear and how it changes behavior. One thing we do know is that the battle is highly lopsided in favor of the wasp. Even when the tarantula does fight back, its fangs are mostly useless, simply sliding off the wasp. Tarantula hawk bodies are hard, smooth, and slippery, have no rough areas, indentations, or ridges, and have rounded bodies. The tarantula has the same problem with its sharp fangs that a person would have attempting to hold a glass beer bottle in one hand while drilling through its side with an electric drill in the other hand: the fangs and drill bit simply slide off sideways. Several observers have reported that, when tarantulas actually attempt to bite and crush a tarantula hawk, loud snapping sounds can be heard as the fangs under immense force abruptly and repeatedly slip off the wasp's body. In the end, the wasp emerges unharmed.[9] Perhaps, rather than fighting, the best spider strategy may be to run and then freeze in hopes the wasp loses interest. I am glad our own species rarely encounters similar situations.

We humans are masters of our lives. We no longer fear large animals that may prey on us, having long since dispatched most of them and their threat. We have conquered many human diseases, though more continue to emerge to challenge us. We have tamed animals and manipulated plants to provide a steady, more reliable food supply. We have made clothes and shelters to make life comfortable. We have made games and toys to entertain ourselves. Tarantula hawks have not mastered their lives as well as humans, although they are a close second. Of course, by "mastered" I do not suggest that tarantula hawks made conscious decisions to alter their lives as humans have (we have no evidence of consciousness in tarantula hawks); rather, nature, through natural selection, has made them masters of their lives. Tarantula hawks live long lives, they have no known predators of adult females, and they can be active any time of day and anyplace they choose. How was this good life achieved? Defense against predators is the most important factor in a long and free life. Without good defenses, animals must either live secretive and restrictive lives or have short lives and try to mate and reproduce before being eaten. No predators successfully prey on healthy female tarantula hawks,[6] although I did once see a particularly small male wasp being eaten by a large praying mantis on a milkweed flower. Pinau Merlin, an Arizona naturalist, reported coming upon a roadrunner—that intrepid predator of many life forms, including rattlesnakes—stealing a paralyzed tarantula from a tarantula hawk and then feeding it to her young. The wasp was left alone. The obvious reason large predators such as roadrunners, other birds, lizards, toads, and mammals don't prey on tarantula hawks is their sting. The sting alone would not be sufficient to protect the wasp from being smashed and eaten by a powerful bird beak or crushing lizard jaws. Here, the second defense, the same defense that protects against the tarantula—the hard, slippery, rounded body shell—provides the necessary time to deliver the sting. The wasp is too tough to be smashed fast enough by beaks and jaws to avoid a sting to the mouth or tongue, and mammalian teeth slip off the wasp body long enough to allow the sting to be engaged. The enormous body size of tarantula hawks relative to most insects and arachnids provides defense against arthropods. If size alone does not do the trick, the sting, hard-body integument, and powerful sharp jaws complete the defense against arthropods.

A universal law of life is that it is always better to avoid a fight with a predator than to actually fight the predator. For a tarantula hawk, why risk losing a leg or an antenna or having a wing crumpled by a bird or lizard if the attack can be avoided? The key to avoiding an attack is communication to the attacker that an attack is risky. Tarantula hawks are masters of communication, using many forms of aposematic warning signaling. Brilliant conspicuous color patterns of reds, yellows, oranges, or whites combined with black are classic examples of warning coloration. Strikingly shiny, reflective, or iridescent dark colors are another example. These patterns tell the predator "see me, I am bright, bold, and dangerous; if you attack me you will suffer." Tarantula hawks with their strikingly reflective orange or shiny black wings and iridescent gunmetal blue-black or black bodies send the warning strongly. To supplement the visual color pattern, tarantula hawks engage in a distinctive jerky movement while they are on the ground, and they flick their wings while moving around, an action that ensures they are seen. Threatened tarantula hawks communicate acoustic warning sounds by buzzing their wings, much as threatened bees raise their buzzing to a high pitch. A final tarantula hawk warning signal is its powerful odor. Humans, as a species with poor olfactory abilities, only notice the odor when massive amounts are released by a threatened wasp. Small amounts of the odor likely are continuously released and operate as a long-distance, early-warning signal to olfactorily cued mammals, warning them not to approach. Given all these warning modalities, no potential predator is left unaware of a tarantula hawk.

Imagine for a moment what freedom from predators means. With no predators, there is less hurry in finding a mate and reproducing; no predator-based reason to have a short, efficient life; no reason to avoid open areas, flowers, or ground surfaces, where predators might take notice; and no reason to limit activity periods to times when risks from predators are minimal. For a tarantula hawk, such freedoms are essential. Tarantulas are not abundant, they are hard to find, they are widely dispersed in the environment, and they are available throughout much of the year. Tarantula hawks require much time and searching for both their own food and for tarantulas for their young. They could not easily pass on their genes to the next generation without a long life and few restrictions on their activities.

The sting has been accepted as an amazing given. Just what makes the sting so powerful? What is the chemistry that makes it so magical? The venom is nearly unique among insect venoms. Most wasp, ant, and bee venoms serve only one role: either offensive in prey capture or defensive against predators. For defense, pain, along with damage or killing power, is important. For offense, pain is irrelevant, perhaps even harmful if it causes unnecessary stress to the prey. Damage or killing power is not beneficial if the prey is to be kept alive and fresh for the young. For offense, the important venom feature is to paralyze, thereby inactivating the prey. Tarantula hawks bridge the gap between offense and defense. Their unusual venom both permanently paralyzes the prey and protects against predators. Pain is the hallmark effect against predators. Tarantula hawk venom damage to predators is trivial; at best, the lethality to mammals is only about 3 percent of that of honey bee venom. Why is tarantula hawk venom nontoxic and nonlethal? Perhaps because natural selection operated against a venom chemistry that is toxic or lethal to tarantulas. A venom toxic to mammals might well also be toxic to tarantulas. Dead tarantulas yield dead tarantula hawk larvae.[8] In addition, tarantula hawks defend no nest and have little reason to damage or kill a predator: the goal is to get the predator to cease and desist, and to open its mouth quickly, allowing the wasp to escape. Only a momentary mouth opening is needed for the wasp to fly away, and the pain alone does that marvelously.

The chemistry responsible for tarantula hawk sting pain is not known. The venom contains the highest-known concentration of citrate, a small 6-carbon polyacidic molecule of any venom, but if, or how, that would cause pain is unclear.[11] The venom also contains the neurotransmitter acetylcholine and kinins, both compounds that can cause pain.[12] These compounds would not cause paralysis of tarantulas, something likely caused by one of a variety of proteins in the venom.[13] Whatever the various active components of tarantula hawk venom, both humans and tarantulas survive a sting; an important difference between the two is that the tarantula succumbs to the tarantula hawk larva, and we do not.

The benign giant. Cicada killer. As ominous as the name sounds, cicada killer wasps are gentle giants among the wasp world. Rather than adopting the strategy of "speak softly and carry a big stick," as Teddy Roosevelt said, cicada killers "speak loudly" and carry a tiny stick. True, they do have a big stick (stinger), and its relevance to cicadas is big, but its relevance to us is tiny. Phil Rau, that intrepid wasp naturalist of the early twentieth century, once wrote that cicada killers "expressed their indignation at being disturbed by the loudest noise that we have ever heard a wasp make."[1] Perhaps this is not surprising, as cicada killers are among the largest wasps in the world, rivaling tarantula hawks in size.

Cicada killers are sphecid wasps in the genus *Sphecius* of the family Crabronidae. The genus *Sphecius* contains five species in the Americas, with four in the United States. As the name implies, they hunt cicadas that are secured in underground cells as food for their larvae. In the purest sense, they are not "killers." They are paralyzers, leaving the "killing" to the larvae as they feed on the paralyzed cicadas. Cicada killers are enormous 2.5- to 5-cm-long solitary wasps, sometimes called "ground hornets," a particularly unappealing and misleading name. They are not hornets, which are best known for their frightening ability to deliver nasty and painful stings, and they do not like most ground, preferring pleasant sandy areas. They are busy wasps active during the warmest times of high summer and bring joy and amazement to anyone fortunate enough to be able to watch them.

The cicada killer life cycle begins in summer in synchrony with the emergence of the adult phase of annual cicadas, which emerge each summer after spending several years underground as immatures. Male cicada killers dig upward from their underground cells, feed on nectar or plant exudates, and establish territories near their emergence sites. About a week later, females begin emerging, again digging directly upward from their overwintering cells. Cicada killers are solitary, gregarious wasps; that is, each female wasp works alone to rear her young; yet, even though the females do not cooperate, they usually aggregate their nests in small areas. Individual nests are often only a meter or less apart. Aggregations range from under a dozen individuals to around a thousand nest burrows in one localized area. Life in the nesting area appears chaotic, with wasps flying every which way and having frequent interactions, none of which are cooperative, with one exception—mating. The brief cooperation between a male and a female to enable their next generation.

Once their single mating is over (females are efficient; why waste time with other males?), females set about feeding themselves with nectar and other sweet liquids, exploring the area, and preparing their nests. Nests are in the form of burrows about 30 to 50 centimeters long and 15 to 25 centimeters deep,[2] dug in the sand with their front legs, likely aided with their mandibles to loosen tough obstacles, and pushed backward up and out of the burrow with the aid of strange enlarged spines, one on each hind leg, called calcaria. Males do no meaningful digging, and, not surprisingly, their calcaria are much smaller than those of females. When the female's burrow is deep enough she shifts modes and focuses on catching cicadas. Our human intuition has led us to expect cicada killers preferentially to locate male cicadas by their loud, boisterous songs. Sound is important in our world, so it is natural we would think this way. In the cicada killer world, sound is of minor, if any, value, and we have no evidence cicada killers can even hear. Maybe hearing in cicada killers would be disadvantageous. Cicada alarm stress calls can reach sound levels as high as 105 decibels, 10 times the loudness of a jackhammer at 16 meters and well over the sustained level of human exposure for hearing loss. Loud cicada stress sounds are known to interfere with mammalian predators of cicadas[3] and might, likewise, interfere with the ability of cicada killers that could hear to capture their noisy prey.

Instead of finding cicadas by sound, cicada killers locate cicadas visually, likely in conjunction with chemicals recognized on contact with the cicada. The female cicada killer visually finds cicadas by slowly scanning up, down, and across the branches of nearby trees where cicadas are located. When a cicada is detected, and for certain recognition, she darts in and out in front of the cicada to provide a better image, much as binocular vision provides better images for humans. She then pounces on the cicada, which often squawks shrilly (if it is a male) and quickly stings it.[2] Paralysis is nearly instantaneous, with the cicada paralyzed within one to two seconds. She then flips the cicada over, holds it belly to her belly with her middle legs, and flies (or attempts to fly) off with the cicada toward her burrow. Cicadas are usually much larger than the cicada killer, making the flight an onerous and limiting task for all but the largest females. Small females often fail in transporting enough cicadas to their nest to successfully reproduce.[4] It pays to be a big female in the cicada killer world. Many times paralyzed cicadas can be seen under trees where they were dropped by wasps unable to carry them.

Joe Coelho, a cicada killer expert at Quincy University in Illinois who also would make a fine helicopter engineer, has spent much of his academic career analyzing just how cicada killers and other wasps can fly with impossibly heavy-appearing loads. The question is somewhat like the famous calculations that bumble bees could not fly; yet they do. Joe found that a cicada killer could just barely fly with a cicada that weighed 1.42 times the wasp's weight. In one population of cicada killers, he found that they solved the problem of carrying slightly overweight cicadas by flying with full lift power and tracking a gently forward and downward gliding path toward the nest. If she hit the ground before reaching the nest, she climbed a nearby tree or tall plant with the cicada and repeated the operation. By this stepwise method, she could eventually reach her nest with her "impossibly large" cargo.[5] Is it any wonder that cicada killers only forage for cicadas within about 100 meters of their nest?

Jon Hastings and Chuck Holliday, respectively, at Northern Kentucky University and Lafayette College, discovered nature had another way of solving the heavy cicada problem. They studied two populations of eastern cicada killers in northern Florida that were about 100 kilometers apart. Both populations had the same four species of cicadas present in about the same ratios. The different cicadas formed a range from small to medium and large. In one location, cicada killers preyed mainly on medium and large cicadas; in the other location, they preyed nearly exclusively on small cicadas. The difference in the wasp populations was obvious: the wasps of the population that preyed on the larger cicadas were much larger in size than the wasps that preyed on small cicadas. The exact causes of this local stabilizing of size differences between the two populations is not clear, although the size of the cicadas fed to their larvae certainly has something to do with it.[6] What is known is that the small wasps simply could not carry large cicadas and large wasps selectively chose large cicadas, even though small cicadas were abundant. Small wasps paid a large foraging penalty because of their small size. By being able to collect only small cicadas, they had to collect twice as many cicadas per young as the wasps in the population with large wasps. What selective force operated on this one population to produce small wasps despite the huge extra foraging cost is mysterious.

After successfully capturing and transporting a cicada to her nest, the female cicada killer places it in the prior excavated cell at the end of the burrow. At this point, she has to make a decision: make a male wasp or make a female wasp. If she chooses to make a son, she lays an unfertilized egg on the cicada, closes the cell, and prepares for the next cell. She can make this decision because she can choose to fertilize the egg or not: fertilized eggs become females, and unfertilized eggs become males. Because males are much smaller than females, that is, about half the size of females, usually one cicada is enough food to produce a male. If she decides to produce a daughter, the female leaves the burrow open (a dangerous proposition inviting parasites, intruders, and thieves into her nest) and seeks another cicada. The second cicada is added to the cell, a female egg is laid, and the cell is sealed. This is the general picture.

Sometimes males get two cicadas and females more than two cicadas; in some populations such as in Florida, or in some species that use small cicadas, as many as four to eight cicadas might be required per cell.[6] After a cell is completed, the female cicada killer uses dirt excavated from a nearby future cell to seal the cell and the tunnel between it and the new cell. She is now ready to hunt the next cicada for her new cell. During her lifetime of a month or so, she produces about 16 cells when conditions are good. Within the cells, the eggs hatch in a day or two, the larva feeds on the cicada(s) for 4 to 10 days, overwinters as a postfeeding larva, pupates the next spring for about 25 to 30 days, and emerges as an adult during the next summer's cicada season.[2]

Sex in the cicada killer world: Most of the chaos in a cicada killer community, as in a human community, revolves around sex. It's the males, not the females, that cause the commotion. Given a choice, females appear to simply want to get mated and then inconspicuously get on with the duties of producing their descendants. Males have only one way to produce descendants and that is to mate with females, a job they attempt with great energy. Males emerge before females and attempt to establish territories in prime areas within the previous year's nesting area. Because at least twice as many males as females are produced, the competition is fierce. If a male succeeds in establishing his small territory around a perching location, the top of a plant, the end of a branch, a rock on the ground, or the bare ground, he must vigorously defend it from other males seeking to take over the territory. Intruders, such as an insect flying by, a small bird, a biologist, or, especially, another male, are investigated. If the intruder is not another male, the perch owner tends to quickly return to his perch. If it is another male, he will attempt to chase it off, frequently head-butting into it. If the intruder does not depart, the two engage in a spiraling dual in which both rapidly circle each other as they ascend into the sky. Serious challenges evolve into grappling bouts in which each male attacks the other, attempting to bite legs, wings, or whatever is available. These grapples can result in both wasps falling to the ground, continuing to grapple, sometimes with a loud buzzing sound heard. Some grapples might be the result of initial mistaken identity in which the male mistakes the intruding male as a female and attempts to secure her. In these territorial competitions, the larger male usually wins. Smaller males can set up territories in marginal areas around the nesting area or can attempt to patrol through the territories of other males in hopes of catching a female first. The smallest males can lurk in vegetation just outside the nesting area in hopes of intercepting a virgin female that somehow emerged and flew through the area to the nearby vegetation.

When a female emerges from her underground cell, she leaves her entrance hole with a characteristic behavior, in which she flies slowly in a straight line toward nearby trees. Mated females fly more quickly, zigzagging or abruptly changing flight, thereby distinguishing themselves from virgins. Males seeing a virgin in flight chase her, attempting to land on her back and fly off together to a resting place where he probes her genital region until locking with her. At this point, he releases his grip and falls backward, often dangling below her and appearing paralyzed or almost dead. Mating lasts on average an hour or so,[7] with the record set in Ruby, Arizona, by a pair I watched for 2 hours 16 minutes. If the pair is disturbed, say, by a potential threat or another male, they fly off in a horizontal tandem with the female pulling and the male providing lift for his own body weight. This idealized scenario for male and female is rarely the case in crowded nesting areas with many enthusiastic males present. Frequently, a mob of several to many males grab onto the female, each male attempting to clasp her, and the ball falls to the ground as a seething mass. Eventually, one male succeeds in engaging her genitalia. He then has the task of pulling her from the mass of other clinging males. In rare, extreme cases as observed by Chuck Holliday, the female or males within the mating scrum might actually die from overheating.

Males with their "live fast, die young" approach to life sometimes are the losers because of their lifestyles. Males live on average 11 to 15 days. Female emergence occurs over a period of 23 to 49 days. Thus, to achieve maximum opportunities to mate, a male is most successful if his short life occurs over the peak emergence period for females. But peak emergence period varies widely from year to year. Some years female peak emergence occurs two to three weeks before or after other years. If a male "guesses" wrong in his emergence time, he might lose, irrespective of his large size. However, a small male might do well if he emerges at the best time.[8] Little wonder a male cicada killer's life is chaotic.

Predators, parasites, and diseases plague cicada killers. Their enormous size, loud buzzing, and bright-yellow and milk chocolate to rufous brown colors make cicada killers conspicuous to predators. Sometimes, this bright warning coloration works well; sometimes, it might not. Western kingbirds in Ruby, Arizona, have specialized in preying on cicada killers. Actually, they prey on the food for the next generation of cicada killers. A kingbird will chase and attack a burdened female returning to her nest with a cicada, forcing her to drop the cicada, which the kingbird grabs and eats. Kingbirds do not attack unburdened cicada killers.

Risks are also present at the cicada killer nest burrow. A variety of flies lurk around the nest entrances hoping to quickly larvaposit (many flies lay tiny live maggots instead of eggs on their targets) on the incoming cicada. If successful, the maggots quickly overcome the rightful owner's egg and devour its cicada(s). More appealing parasites, at least to people, are the colorful velvet ants. These large velvet ants, sometimes called cow killers, sport beautiful fur coats with patterns of red and black, orange or yellow and black, or yellow or white. The velvet ants associated with cicada killers are among the world's largest velvet ants, largely because their young have enormous nutrition available in the form of the larva of cicada killers. Once again, you are a reflection of what you eat.

The last, but not necessarily least, of a cicada killer's risks is another cicada killer. Cicada killers leave their nest entrance burrows open, even when a cicada is present in a cell. Other females can exploit this opportunity to take over the cicada or even the entire burrow from the absent female. Such a takeover by a usurping intruder, if successful, is far easier than digging a new burrow and catching cicadas. Through a clever series of experiments using trap nests, Chuck Holliday and his associates showed that this apparent "kleptoparasitism" can reach more than 50 percent in experimentally abandoned nests.[9]

What about the sting? Surely, in a wasp this large with a 7-mm stinger, the sting must be of use in defense. Yet, few, if any, reports of attempted predation on adults, much less successful predations, are reported, which suggests that the sting is not needed. Perhaps simply the wasp's appearance as a supersized yellowjacket, a beast to be feared and avoided, is all that is needed. The fact that few people seem to be stung by cicada killers could also suggest that the defensive value of their sting is minor, so minor that a person must exhibit real talent to be stung by a cicada killer. In many years of studying cicada killer wasps and their venom, I was never stung. During this time, wherever I went and mentioned cicada killers, people expressed great fear or apprehension of cicada killers. I was always asked how much does their sting hurt. My answer: "I've never been stung, but I expect it would not be too painful" somehow seemed unsatisfying to both the asker and to myself. I was the expert. So why such an unsatisfactory answer? Eventually, this got to me, and I had to do something. What to do? Ah, ask Joe Coelho, the expert mentioned earlier! Joe answered, "Oh, it is pretty trivial, sort of like a pin prick, not much pain." My theory that it wouldn't hurt much was supported. But still, maybe Joe was just underplaying it. Next, I looked in the literature and found one report from 1943. Charles Dambach was stung by a "large specimen" near the tip of his right index finger and wrote, "An initial sharp pain was followed by numbness, a slight swelling and stiffness which lasted about a week."[2] Again, this supported the theory that cicada killer stings wouldn't hurt that much (note, no superlatives in his quote). Finally, I realized that I had to get the real story myself. In the popular media and in academic circles, the modern legend has it that "Schmidt is the guy who likes to sting himself with any imaginable stinging insect." Cicada killer wasps are a main source for this legend. Yes, I needed to get the data on the pain level of a cicada killer sting. No, I hadn't been stung in the heat of the battle. And, no, I did not want to sting myself. What to do? One day opportunity struck; a western cicada killer (Joe was stung by an eastern cicada killer) happened to be sipping nectar from a flower, and I was missing my net. I grabbed the wasp and wham, or perhaps I should say "slap," I got stung. It wasn't like getting hit by a bullet or a flaming torch; rather, it was like being stuck in the palm by a thumbtack. The pain was sharp and immediate, lacked any burning sensation, and lingered for about 5 minutes. There was no swelling, and the pain was entirely gone in 20 minutes. Pain level 1.5 on the pain scale, a lot less than a honey bee sting: a low level of pain for such a large wasp with such a huge stinger. Theory confirmed in person.

If cicada killers don't attack us, and don't sting us, and their stings don't really even hurt much, why then have they become so imbued in human popular media? I ask for forgiveness from the medical and pest control industries here, for I associate them with popular media. When a local medical treatment for insect stings is to be marketed, does the industry feature one of the true culprits, a honey bee or yellowjacket wasp? No, they often feature an enormous picture of a cicada killer, flying from blackness right at you. When the pest control industry, the guys in clean, starched white outfits, who come to your house to spray or treat cockroaches or termites, need to display an insect for their magazine feature on stinging insects, do they feature a honey bee or yellowjacket? No, they feature a huge picture of a cicada killer, again with a black background. Why do professionals who know, or should know, that cicada killers are harmless, prominently feature cicada killers to represent stinging insects? The short answer is human psychology. Bigger is scarier. By simply mimicking the smaller, dangerous stinging yellowjacket wasp, the cicada killer becomes in our mind a "huge yellowjacket," thereby winning the war of wits with humans—without ever having to show its sting. This same mimicry of stinging yellowjackets also appears to work on other large animals. Having nasty little fellows around that look like you has its benefits.

Mud dauber, the world's most familiar solitary wasp, adorns our structures with her graceful nest of mud placed on walls, under roofs, and, in bygone days, inside our outhouses. The mud dauber, *Sceliphron caementarium*, is also sometimes called the "dirt dauber," or, if one wishes to be formal, the "black and yellow mud dauber." She is the only wasp I know to have a book dedicated to her, and not just dedicated to the species but to a single individual known as Crumple Wing. Arnold Menke, a distinguished wasp expert from the Smithsonian Institution, also succumbed to the spell of mud dauber wasps. Arnold, coauthor of *Sphecid Wasps of the World*, a 600-page bible of sphecid wasps commonly called the Big Blue Book, adopted the pen name of "Mud D'aub" for many of his popular writings on wasps. I suspect Mud D'aub will not become as familiar a writer as Mark Twain, another pen-named author, but they both share the distinction of choosing their names based on their work and love: in Menke's case, the familiar and beloved mud dauber wasp; in Clemens's case, the riverboat terminology "mark twain," for two fathoms of water depth, an indication of safe navigating water in the Mississippi River.

For all their familiarity to humanity, mud daubers seem cloaked in superstition and misinformation. Many people, especially in the southern regions of the United States, fear these wasps, especially their stings. Lynne Bachleda in her book on dangerous wildlife, wrote "mud daubers can sting painfully" and "the mud dauber's sting is potentially lethal to those highly allergic and prone to anaphylactic shock."[1] Rod O'Connor writes, "[the mud dauber sting produces an] unusually mild immediate reaction on humans, i.e. negligible pain and swelling," followed by, "but an authenticated case of mud-dauber sting death has since been discovered."[2] The "authenticated case" turns out to be an unpublished private communication, a perfect source for starting (or continuing) an urban legend. Not to be left out in the vilification of mud dauber wasps, Dr. Claude Frazier, a famous physician from North Carolina, showed in his review of allergic reactions to insect stings a photograph of a mud dauber along with the usual suspects: honey bees, yellowjackets, baldfaced hornets, paper wasps, and bumble bees. He also featured a cicada killer and a velvet ant. In fairness to Dr. Frazier, he never claimed any of these solitary wasps actually caused allergic reactions, but their guilt is implied by association.[3] The record reveals not one allergic death documented from the sting of a mud dauber; indeed, even getting stung once, much less twice or more as needed for an allergic reaction, requires an exceptional talent.

More favorable views of the familiar mud dauber wasp trace to the famous early American naturalist John Bartram, who penned in 1745 the first observation of paralysis instead of death in a stung prey of a solitary wasp. He recorded this observation with the mud dauber: "Only in some manner disable the spiders, but do not kill them . . . that they may be preserved alive and fresh until the egg hatches, which is soon."[4] Bartram also noted that during their "Labours, they make a very particular musical Noise, the sound of which may be heard at ten yards distance." Bartram's first recorded observations are as correct today as during his time.

Perhaps the most unabashed fan of mud daubers was George Shafer, a professor of physiology at Stanford University. George investigated the basic physiology of digestion, plus the life history and biology of mud daubers. His affection for the wasps was obvious in his writings, with the invitation "To the prospective reader, I wish to commend this royal, thread-waisted wasp, *Sceliphron caementarium,* possessed of such grace and seeming personality as to invite acquaintance."[5] His writing entertained and inspired many young naturalists, including me.

The scientific lore of mud daubers is as rich and diverse as its folklore. Mud daubers become musical masons as they work the mud from which they build their nests. The "musical noise," to quote Bartram, is generated by contracting the wasp's thoracic flight muscles, thereby vibrating the head and mandibles while emitting a high-pitched sound. When digging mud from the source and then plastering it on the nest, the wasp uses sounds of different frequencies, apparently optimizing the digging, plastering, and smoothing processes.[6] The finished mud nest sometimes then becomes a preferred attachment site for barn swallows, birds that also fashion their nests from mud.[7] Mud dauber nests sometimes even become the targets of downy woodpeckers, which chisel holes in the nests and extract a meal of cell contents.[8]

Mud daubers are not simply fine masons (their scientific species name *caementarium* is derived from Latin for "mason"), they are chemists as well. Within mandibular glands in their heads mud daubers produce geranyl acetate, a compound with a pleasant floral or fruity rose aroma to the human nose, and 2-decen-1-ol, which has a fatty odor.[9] The purpose of these odors in unknown, but likely they serve as chemical warnings or defenses directed toward predators. Other mud dauber talents include flower pollination, in which they rate the 10th-most important pollinator of carrots in Utah.[10] They are about average among insects in their ability to survive gamma radiation produced by cobalt-60. The American cockroach, the big juicy cockroach that emerges from sewer drains and is loathed by homeowners, was also part of the study. Perhaps surprisingly, the cockroach was the most susceptible to radiation of all tested insects,[11] which refutes the legend that after a nuclear war the only survivors would be cockroaches.

A final superlative talent of mud dauber wasps is their ability to invade and colonize new lands. No other solitary wasp can match their dispersal ability. Its closest competition among bees might be the honey bee, but its dispersal was almost entirely with the help of humankind who intentionally brought them to all peopled major landmasses. Mud daubers are not intentionally spread by humans; nevertheless, they have spread, among other places, to Europe, Japan, and even the Galápagos Islands made famous by Charles Darwin. Their dispersal seems to be as mud nests attached to shipping boxes and incidentally transported in commerce. Once in a new location, mud daubers appear to have good success in colonizing. Curiously, the first reported records of mud daubers in France and Japan were in 1945, about the time World War II ended and reconstruction in Europe and Japan began with American materials shipped from North America, the native home of mud daubers.

What makes mud daubers so omnipresent and successful? Part of the answer is their natural history. Mud daubers build their nest of mud and provision them with spiders for their young. Both mud and spiders are universal components of most habitats. Mud daubers are only semi-fussy about their spider prey. Their favorites are orb weaver spiders, followed by crab spiders and jumping spiders.[12] Mud daubers locate spiders visually and pounce on them. Spiders are recognized by cuticular components on their exoskeleton. If the attacked object is not a spider, the wasp discontinues the attack and continues searching. Recognition cues are distinctive not only to spiders but also to specific types of spiders. In a study, Divya Uma extracted the waxy components from the integuments of several types of spiders and applied them to paper dummies. Mud daubers attacked the dummies coated with extracts of spiders that made two-dimensional webs (familiar orb weavers found in gardens) but avoided dummies with extracts of spiders that make three-dimensional webs, in this case the common gray house spider, a cobweb spider that makes a tangled web. The wasps stung the two-dimensional web makers or dummies coated with their extracts but rarely attempted to sting the three-dimensional web makers or dummies coated with their extracts.[13] One species of jumping spider, a normally accepted group, evolutionarily "out-foxed" the mud dauber by having a different chemical coat that was not recognized as "spider" by the wasp. The disguise was even better—the spider looked like a carpenter ant.[14]

When the mud dauber finds a suitable spider (a test best failed in this case if the spider is to survive), the wasp grabs it with its mandibles and front legs and stings the spider under the cephalothorax, the head-thorax unit. Three stings directed toward the nerve ganglia that control jaw and leg movement are the usual pattern. The stung spider immediately becomes limp and paralyzed. The mud dauber will often press its mouth to the spider's mouth and drink from it. Sometimes, the wasp will also chew through the spider's leg base or abdomen and drink fluids. Why the wasp does this is unclear. By drinking the spider's fluids, the mud dauber deprives its young of a high-quality spider, and sometimes the spider entirely, as the spider may be discarded after the drink. Perhaps the reason for drinking from the spider is to obtain valuable protein, a nutrient woefully lacking in the mud dauber's usual diet of sweet nectar. When not used solely for personal benefit for the wasp, the spider is transported to the mud nest and pushed into a premade mud cell. An egg is laid on the first spider and the wasp goes in search of more spiders. When 6–15 spiders have been added, the cell is capped with mud and a new cell started, attached to the growing mud blob. A dozen or so cells are constructed, filled with spiders, and sealed in this fashion during the female's six-week to three-month life span. The egg in the cell hatches into a tiny, nearly transparent larva, which sets about feeding on the provided spiders. When all spiders are consumed, the fattened larva spins a silk cocoon, rests several days while the connection between the gut and rectum forms, and then defecates in the bottom of the cell. The larva, now called a prepupa, an intermediate stage between the true growing larva and the pupa, rests quietly overwinter in its silken enclave. At the end of winter, the prepupa molts into the pupa, which then metamorphoses into the adult. Adults rest in their cells several days while hardening their cuticle and then chew through the hard mud cap to emerge. Males, like males of most sphecid wasps, are smaller than the females and emerge shortly after the first females in the early season. Though mating behavior in mud daubers has been a neglected topic, females apparently mate shortly after emergence and start their summer nest building and foraging behavior.

Mud daubers can and do sting spiders. Do they sting in defense? Perhaps, but it's unclear at best. If a mud dauber is grabbed, it will curl its abdomen against the offender in a stinging motion, an action males, which have no stinger, and females perform equally. When this happens, the reaction of most people, entomologists included, is to release the wasp immediately. Thus, no sting is received and the wasp wins. Is this all bluff, or is there real substance behind the action? I support the bluff concept because males are released as readily as females, and both males and females are mimicking honey bees and yellowjackets that *do* readily sting. Is it worth taking the risk that one's identification is wrong? Evidence against meaningful pain subsequent to the "stinging" action is the infrequency of documented stings to humans. I personally know of nobody who has been stung. Negative evidence is, of course, only a teaser; I fully accept the idea that mud daubers are able to sting people, as illustrated by Rod O'Connor's earlier statement of a sting that produces "negligible pain and swelling."

Analysis of mud dauber venom can also provide cues to its defensive effectiveness. Stings from insects with effective defensive venoms hurt, cause damage, or both. Painful components include basic, that is alkaline, peptides, often with the small neurotransmitters such as histamine, acetylcholine, and serotonin. Common pain-inducing peptides among wasps are analogues of bradykinin, a small peptide that acts on the heart and causes intense pain. Mud dauber wasp venom lacks all of these components.[2,15] A total lack of toxicity data to mammals or to arthropods hints that mud dauber venom also lacks meaningful toxicity. Stung spiders, though powerfully paralyzed, exhibit minimal indications of toxicity, and the heart and likely the digestive system and blood cells are unaffected. This implies no direct toxic or tissue-damaging effects are present.[16]

Once again, we reach an impasse. Are mud daubers dangerous? Do their stings hurt? We have little evidence concerning the pain, let alone potential danger, of a mud dauber sting. To make matters worse, I have never had the (mis)fortune to be stung in the process of normal fieldwork or research by a mud dauber (not that I haven't been casual and lackadaisical around them). Seemed to be déjà vu cicada killers. I hadn't been stung, predicted they wouldn't hurt much (or they would have already stung me), and needed some facts. Only this time I was lacking a Joe Coelho to advise me on the sting: I had, at best, the tenuous comments written by Rod O'Connor more than a half-century ago. OK, time to bite the bullet—grab a mud dauber, get it over with, and go home with a data point and new wisdom.

It was a fine June day in Willcox, Arizona. No rain had fallen in months, and the only water around was in stock tanks for cattle. I was visiting a stock tank filled by an Aermotor windmill pumping water from underground into a large metal trough formed from half an old underground fuel tank excavated from a gas station. Fortunately, for me and the mud daubers, the check valve to the intake had failed, allowing the tank to overflow, forming a big puddle of mud. Many mud daubers were busily collecting mud from the puddle edges. Opportunity. I grabbed a big mud dauber and guided her posterior to my left forearm. With some struggling, she planted her stinger in my flesh and delivered her payload, after which I let her go, and she flew off. Underwhelming. That's the only word to describe the pain. A minor, immediate sharp pain was detected. The pain did not jump out and grab my attention. Instead, it produced a pain somewhere between 0 and 1 on the pain scale. Shortly thereafter, a minor burning developed, earning 1 on the scale. This pain soon dissipated, leaving no visible swelling, redness, or sign of a sting. A trivially minor pain for a data point of major importance. Time to flee the heat of the day, have a cold beer, and relax writing notes.

THERE ARE TOO MANY SOLITARY WASPS to study them all. However, the large iridescent blue digger wasp, *Chlorion cyaneum*, commands our attention and respect, if not outright fear. Perhaps this wasp is intimidating because of its large size, 25–30 millimeters long; perhaps it is intimidating because of its flashy, iridescent brilliant blue body and purplish-black wings; perhaps it is intimidating because of its narrow thread-waist, something people seem to fear; or perhaps it is intimidating simply because of its rapid, jerky movement. Or maybe we are intimidated by all of the above.

Who is this colorful wasp who so catches our attention and had a 1970 El Salvador 30-cent stamp honoring her? She is a member of a small, elite genus of 18 thread-waisted wasps, most of which live in the Old World. The few studies of the group indicated that they are specialists on crickets, which they sting causing only transient paralysis in some species and long-lasting paralysis in others.[1] One widespread African species, *Chlorion maxillosum*, takes parental care to a minimalistic extreme. Not only does mother wasp fail to dig a burrow to house her young and their cricket food, she doesn't even transport the cricket to a safe location. Instead, she stings the cricket into short-lived paralysis, lays an egg on the cricket, and abandons it to fend for itself. The cricket soon recovers from its paralysis and either returns to its own burrow or digs a new burrow. The *Chlorion* larva now feeds safely and securely on its cricket in its ready-made cricket burrow.

Another species from North Africa provides slightly more parental care. She drives the cricket from its burrow, stings it into transient paralysis, lays an egg on the cricket, then drags the cricket back into its burrow and seals the burrow. North America has only three species of *Chlorion*; the two studied species both sting their prey into enduring paralysis. The common blue cricket killer digs legitimate 6- to 44-centimeter-long burrows, usually in sandy soil, and hunts crickets in the local area. Crickets are seized, stung in the underside of the thorax into total paralysis, and dragged back into the pre-excavated burrow. This is an ordinary enough behavior for solitary hunting wasps, but the blue cricket killer adds a unique twist. She often digs her burrows deep within the burrows of cicada killers. If wasps can be lazy, we might call this laziness; otherwise, we could call it efficiency, or safety and security disguised within another's home. In any case, the cicada killer doesn't seem to pay attention to the intruder and both coexist peacefully.[2]

Chlorion cyaneum, let's call her the iridescent cockroach hunter, is the most unusual member of her group. She is particularly fond of sand dunes and other sandy areas, and she refuses crickets. She also can withstand extreme sand temperatures up to 50°C (122°F).[3] What makes her so unique among her 17 relatives is her choice of cockroaches—sand, or sand-swimming, cockroaches. These cockroaches literally dig and swim through loose sand, and the females are wingless. The males are winged, flattened, tan cockroaches frequently attracted to lights at night. The iridescent cockroach hunter digs a burrow 15–30 centimeters long and provisions the cells with female, immature, and male sand cockroaches, which she stings into total paralysis.[1]

I became interested in iridescent cockroach hunters when I saw them in the field, strutting around flipping their wings, almost saying "see me, here I am" with the implied message "better leave me alone." Wow, this wasp seems to be telling me something; but is it all bluff like a gopher snake hissing and shaking its tail in the leaves to imitate a rattlesnake, or is it real? Oh, these solitary sphecid wasps are getting tiring. So many seem intimidating, yet none follow through with stings. Worst of all, records of stings are rare to nonexistent in most and totally lacking for *Chlorion*. The best we have is Eric Eaton's comment on his blog, *Bug Eric*, that one cannot easily examine live specimens "without getting painfully stung" (bugeric.blogspot.com, August 18, 2010). OK, time to get it over with. The wasp is not going to voluntarily sting me or anyone else, so I reached into the insect net and grabbed a fine female, who stung my fingertips twice during the removal process, and applied her to my right forearm. The pain was sharp, with a nettle-like flare. Anyone who has walked through an eastern North America nettle patch knows the feeling. The pain, fortunately much less than real nettle stings, lasted about 3–5 minutes before the last rash prickles disappeared. Pain rating of 1+ on the pain scale, more than a mud dauber, but certainly less than a honey bee sting. Another ordeal survived.

Paper wasps are well-known, pain-causing, stinging social wasps that build open papier-mâché nests attached under roof, door gable, or other sheltered places. Nobody, or at least nobody with intact nerves, doubts the painfulness of their stings. They live in social colonies with overlapping generations, task specialization, and parental care. They also have solitary relatives in the same family, the Vespidae, that look similar and prey mainly on caterpillars. Paper wasps evolved from a lineage of these solitary wasps, thereby providing possibilities of looking into the origin of the paper wasps' painful defensive venoms. Did the paper wasp ancestor already painfully sting, or did painful stinging evolve after paper wasps branched from their solitary relatives? Fortunately, many solitary relatives of paper wasps exist and provide a convenient means to probe this chicken-or-egg question.

Meet the "walk on water" wasp, sometimes referred to as the "Jesus Christ" wasp. Actually, several wasps, especially in the tongue-twistingly named genus *Euodynerus*, routinely land on water to take a drink. But here the main focus will be on the species *crypticus* that especially resembles some paper wasps. These wasps do not literally walk on water; instead, they come to open-water surfaces like mini-helicopters lowering themselves from the sky, landing gently on the water with legs widespread and wings held obliquely back and raised as if ready to take off at any moment. Without moving or walking, they take a deep drink from the water surface for 12–15 seconds, after which, like heavily laden tanker helicopters on their way to fight a forest fire, lift off slowly from the water surface and fly away.

This behavior raises several questions: Why do the wasps perform this behavior of landing on open water, risking drowning? Why do they want so much water? The first question is harder to answer with certainty but appears to be a way to reduce the risk of predation. In natural situations, the wasps are rarely found drowned under the surface. Perhaps the risk in these normal situations is not that great, unlike in the artificial situations of man-made swimming pools where they are sometimes found drowned, especially after active kids cannonball into the water. By landing on water, the wasps avoid ambush by various predators, especially frogs, lurking near the water's edge.

The answer to the question of the great need for water by water-landing wasps comes from their life history. Dwight Isely in 1913 described in detail the natural history of *E. crypticus* in Kansas.[1] The female wasp selects very hard, dry surfaces in bare ground to dig her burrow. Such hot, dry areas likely have fewer intruders, predators, or parasites than other areas, but the ground is rock hard. The wasp solves the problem by moistening the soil and then removing large mud balls dropped a short distance away, littering the bare area around the burrow. Her vertical nest goes down about 10 centimeters and has one or two cells. Isely described one wasp making 16 water trips and removing 86 clods of earth in 40 minutes while excavating her nest. When the digging is finished, the wasp hunts skipper butterfly caterpillars that she extracts from their tough silken retreats within crumpled leaves and stings each caterpillar three or four times in the throat and toward the nerve ganglia that control the legs and jaws. Five to seven mostly paralyzed caterpillars are stored in a tangled mass in a cell, an egg laid, and the cell closed. In Arizona, the yellow paper wasp, *Polistes flavus,* also floats on the water's surface while collecting water. It also looks remarkably similar to *crypticus*. The main difference is that *crypticus* is stockier. Is this a case of mimicry or simply an ancestral and a derived species that look similar?

Back to the question of sting pain: did sting painfulness or the sociality of paper wasps come first? Two points are noteworthy. First, the social wasp nest is exposed and vulnerable to all sort of predators, especially large predators, whereas solitary *crypticus* has very little to defend, especially from large predators unlikely to dig into rock-hard ground for such a small reward. Second, *crypticus* needs to keep its caterpillars alive, fresh, and paralyzed. *Polistes* kills its caterpillars by chewing them into meat balls to feed immediately to the young. Intuition would say that *crypticus* has little need for painful or damaging venom active on large predators and, to the contrary, damaging venom might be disadvantageous by killing the prey, causing it to spoil. In contrast, *Polistes* has no need to preserve prey alive but does have great need to have painful and damaging venom to deter predators. Thus, we would predict that selection pressure would cause pain to form during the process of social evolution in *Polistes,* not before in the purely solitary phase. The chicken comes before the egg.

Testing time. Once again *crypticus* and other solitary relatives of paper wasps are unlikely to ever sting anybody voluntarily or in defense. Ready. Grit your teeth. Let's go. Back to the stock tank, only this time the wasps were collected while floating on the greenish water. Three different *crypticus* were placed on my arm and enticed to sting. Each produced a low-level burning, somewhat analogous to a microbit of paper wasp venom. Being charitable, the pain rated at best a 1 on the pain scale. Not content with just one species of eumenid wasp (the solitary relatives of paper wasps), I captured a large potter wasp in Shepherd's Tree caravan park near Ellisras in South Africa and had her sting my wrist (she wouldn't/couldn't sting through my fingertips).

Again, at best a 1 on the pain scale. I was content with these two experiences, but fate apparently was not. One day while I was walking sandal-clad through a mesquite flat, I felt a sharp pain under the middle toe of my left foot. It was sharp and somewhat itchy, but not burning like a paper wasp sting. This one rated higher, a 1.5 on the pain scale. The culprit turned out to be a yellow eumenid wasp. Looks like solitary wasps, either eumenids or sphecids, are not able to produce meaningfully painful stings.

My son Kalyan, who had just turned 8, asked, "Dad, are there any insect tanks?" "Well, if you mean hard as an army tank, fast as an army tank, and with the firepower of an army tank, then, yes. They are called velvet ants." Velvet ants? What are velvet ants? No, they are not ants, they are wasps that look like sturdy ants, often covered in dense velvety red, orange, yellow, white, or black fur, hence their name. Unlike ants, they do not live in social colonies, have no queens, and live strictly solitary lives. Commonly recognized velvet ants are female wasps that have no wings, nor even a hint of wings. They can (and we might add, readily do) sting and have the distinction of possessing the longest, most agile stinger found within the insects. Female velvet ants literally are micro-tanks on six strong, short legs. They are hard as a rock, sometimes so hard that entomologists bend the steel pins used to mount insects for study. For young children, who are always up for a challenge, one of my favorite challenges is to spot a velvet ant running on the ground and say, "I'll bet that you can't smash that insect." Fighting words. The universal response is to stomp on the velvet ant. It simply makes a velvet ant impression in the ground; the velvet ant picks itself up and runs off. Stomp, stomp, stomp. The same result. But don't try this barefoot.

Unlike the females, male velvet ants do not look at all like ants or their sisters. Instead, they sport fine functional wings, are usually black or brown, sometimes with splashes of color, and appear more like slow, furry, ill-defined flying insects. Although they are wasps, they don't have that sleek, agile appearance of most wasps, resembling mini-flying, meandering teddy bears. "Teddy bears" they are. They cannot sting like females, cannot really bite, and when captured sing and release perfume. Quite cute and harmless. Females, too, are cute, but far from harmless.

In 1758, Carl Linnaeus, the father of modern taxonomy, described several species of velvet ants, among them *Mutilla europaea*. This uncommon velvet ant is also one of the most unusual. Of the roughly 6,000 velvet ant species, only this species and perhaps one or two other closely related species are known to use highly social insects as hosts; all the rest of known hosts are solitary insects or, at most, primitively social. The habit of parasitizing bumble bee and honey bee colonies appears to be the reason *M. europaea* attracted early attention. In the eighteenth century, sugar and sweets were expensive and in scarce supply, and honey was highly prized for a variety of reasons in addition to its sweetness. No wonder something that attacked honey bees was noted and described early. The invading *M. europaea* targets mainly colonies of numerous different species of bumble bees. Struggles between guarding bumble bees and the velvet ant are rare, perhaps good for the bees, as attacking bees tend to end up dead. Once in, the velvet ant makes itself at home, moving about unimpeded by bees and parasitizing bee postfeeding larvae and pupae within their silken cocoons. The velvet ant egg hatches into a larva that feeds as an ectoparasite on its bee host, molting four times as it grows, and finally, having finished its food supply, spins its own cocoon within the bee's cocoon, pupates, and emerges as a fresh adult 30 days after the egg was laid. These velvet ants can be serious threats to bumble bees with as many as 76 velvet ants produced in one colony.[1]

Mutilla europaea will on occasion enter honey bee colonies, thereby ensuring its fame. Some honey bees, as with bumble bees, attack invading velvet ants, suffering the same lethal outcome. After a few minutes, the honey bees avoid the intruder, who seeks bee larvae spinning their cocoons and lays an egg inside with the spinning larva. Several literature reports described scary scenarios of bee losses within beekeeper colonies, something, which if accurate, seems to have disappeared with modern beekeeping, and likely was greatly exaggerated at the time. These stories fit nicely with an enduring human tendency to embellish stories to make them more interesting. Not to be outdone by the Europeans, a variety of late nineteenth- and early twentieth-century American writers described dire situations, such as velvet ants being "serious insect enemies of honey bees," statements best regarded as fanciful or old wives' tales (all written by men), for no valid documented reports of invasion and damage to honey bees in the New World are known.[1]

The most famous North American velvet ant is the cow killer, so named because those stung by one felt the sting could "kill a cow." This velvet ant, *Dasymutilla occidentalis*, vies for the most attractive species, with its short, neatly aligned, red and black velvety coat and two pleasingly arcing, large red spots on its abdomen. This beauty is memorable to anyone who sees it and adorns most nature guides that cover insects. The cow killer was prominently figured in 1703 by James Petiver, making *D. occidentalis* the first North American velvet ant species accorded such recognition. Misinformation surrounds any conspicuous organism, and such is the case with the cow killer. C. V. Riley, the first official U.S. government entomologist and first Smithsonian Institution curator of insects, published in 1870 a letter from a Texas man who reported a cow killer entering a beehive and killing bees that attacked her.[2] This began the vilification of the cow killer, which lasted until 1932.

Velvet ant life history shares many similarities, and some differences, with other solitary wasps. Female velvet ants actively search for hosts to parasitize. In this search, they are both flexible and eclectic: flexible in accepting a number of different host species; eclectic in only accepting host postfeeding larvae or early pupae. Cells that contain host eggs, feeding larvae, maturing pupae, or simply host provisions are rejected. One other stringent requirement is that the host must be housed in some "package," usually a cocoon or hard shell, as in a fly puparium or beetle pupation case. Most hosts are solitary wasps or bees, rarely other diverse hosts, including puparia of tsetse and other flies, pupae of moths within hard cocoons, beetle pupae within hard cases, and cockroach eggs in hard oothecal cases. Once a suitable host in the right stage is located, the female chews a small hole in the cocoon or package, inserts her sting apparently to sense the conditions within the cocoon, and lays an egg. In most situations, the female appears not to sting the larva or pupa, though she might sting some pupae to arrest their development.[3] Once the egg is laid, she closes the hole in the cell with nearby nest materials cemented with saliva and resumes her search for more hosts. The egg hatches in two to three days, the larva feeds on the resting host, molting as it grows, finishes eating all of the host, spins a cocoon, defecates, molts into a pupa, and finally metamorphoses into an adult. During warm seasons, the cycle is uninterrupted. If winter is approaching, the young velvet ant overwinters as a postdefecating prepupa that then becomes a pupa in the spring and continues the cycle. Until recently, the rule of only one velvet ant young per host cell was apparently strictly followed. But nature, and in this case nature in Australia, tends to make exceptions to rules. Two Australian species of velvet ants produced four young per cell of their mud-nesting wasp hosts.[4]

One important part of the life cycle remains: courtship and mating. In many velvet ant species, sex resembles a task to dispense with as quickly as possible before getting on with life. Males fly over promising areas, searching primarily by odor for virgin females. Virgin females release a sex pheromone attractant that the male detects while flying overhead, inducing him to drop to the ground where he frantically searches for her. Vision plays little or no role, as he often runs right past her without noticing. Once he stumbles into her, he immediately recognizes her by contact chemicals and mounts her while singing with his abdominal stridulatory organs and buzz-honking with his wings. He probes her abdomen tip with his genitalia. If receptive, the female extrudes her sting a remarkably long distance and opens her terminal abdominal plates, allowing him to engage his genitalia. Now mating in the fast lane begins and lasts only about 15 seconds. The female runs off never to mate again. The abandoned male is on his own to search for more females.

This mating story is not universal among velvet ants. Velvet ants have the problem that females can't fly, and their ability to disperse is limited by how far they can go on foot. In many velvet ant groups, this problem is reduced by the male grasping the female and flying with her as they mate. He often carries her for 2 hours, mating with her for about a minute each of five times, and then depositing her in a new location.[5] The new location might be across a stream or other physical barrier she could not cross on her own. To fly while carrying a female, larger size is favored in the male. Denis Brothers, a talented South African entomologist with an irrepressible enthusiasm for velvet ants and other fascinating wasps, shows a picture of a mating pair of velvet ants in which the male is nearly three times as long and, by my calculations, 25 times the weight of the female.[6] To put this in human perspective, this is like a 120-pound woman dating a 3,000-pound man. One suspects the male velvet ant would have no trouble flying with his lady.

Why are velvet ants insect tanks? Why do they need such armor, such firepower, such speed? Defense. Defense against what? Nearly everything: hosts resisting their attempts to parasitize; competitors, including ants, at sources of nectar or honeydew; and a nearly endless number of predators looking for lunch. In response, velvet ants evolved the best and strongest defenses known in insects. Most insects have only one or two defenses to augment behavior and lifestyle. Among these defenses are cryptic camouflage to prevent detection, strong jumping legs to bound away, powerful wings to rapidly escape, hard shells to resist puncture, internal poisons to make them unpalatable, chemical defenses to repel attacks, and stings. In addition to behavior and lifestyle, velvet ants have an amazing six defensive systems: their sting; their rock-hard body; their short, powerful legs for rapid escape and wresting free from grips; their aposematic warning colors; their warning, defensive sound; and their warning, defensive chemicals. Not all velvet ant species have all six defenses. For example, nocturnally active velvet ants lack, and have no need for, warning colors. One might reasonably ask why velvet ants need so many defenses when other insects get by with one or a few? Clues come from velvet ant life history. Hosts for velvet ants are frequently present in low numbers, are widely dispersed, and often live in open, exposed areas such as sandy patches or dunes, where avoidance of detection is nearly impossible. Velvet ants have generally low numbers of offspring, making survival of parent and offspring crucial. Velvet ant females cannot fly to escape. Finally, velvet ant females have very long life spans, often well over a year. Combined, these life history traits mean velvet ants live for a long time, nearly constantly exposed to a variety of predaceous spiders, beetles, ants and other insects, lizards, birds, mammals, and even toads. About the only predators they are not exposed to are fish. Velvet ants need effective defenses for each of these types of predators.

The most-studied model of velvet ant defenses is our old friend and poster child of velvet ants, the cow killer. The ultimate primary velvet ant defense is the sting. A cow killer's sting sets not only the record for sting length relative to body length among aculeate Hymenoptera, it also is the most flexible and maneuverable of the stings, able to reach all parts of the body except very narrow portions of the thorax and abdomen. This length and maneuverability is achieved by having the stinger looped forward within the abdomen, then around, and finally back to the tip of the abdomen, much as a watch spring of an old watch is coiled within its case. At the exit point in the abdomen are muscles and plates that guide the stinger toward the right, left, or down and forward. As far as we know, the sting is only rarely used for stinging prey (which are already in a nearly immobile resting phase) and essentially solely for defense. The defensive value of the sting is especially important against the relatively enormous predaceous birds, lizards, mammals, and toads, but it also works well against spiders, praying mantises, and other smaller predators.

The sting alone would be poorly effective against a crushing blitzkrieg attack by a bird or a lizard. This is where the second primary defense comes in. Just as an army tank cannot be easily crushed, bird bills, lizard jaws, and mammal teeth cannot easily crush or pierce the shell of a velvet ant. The force to crush a cow killer was more than 11 times that required for a honey bee.[7] This figure is only part of the story. The entire body of the velvet ant is rounded with tight-fitting body parts and no soft, membranous gaps where teeth or fangs can get through. The result is that bills, jaws, teeth, and fangs slip off like chopsticks sliding off a greased marble. No grip, no crush. In the meantime, while the hard shell is keeping the velvet ant intact, the sting is brought into play. The instantaneous result is the bills, jaws, mouth, and fangs are snapped open allowing the velvet ant to escape unharmed while the predator rubs its mouth or plows it through the sand, trying to remove the pain.

A female velvet ant thorax is an amazing box of muscles. Since she doesn't have wings and can't fly, the space normally allotted to wing muscles is taken up by huge muscles to the legs. This gives her some of the strongest legs in the insect world, legs perfect for both wresting her body free from a predator's grip, and for running rapidly once free. The powerful legs in conjunction with the hard body combine to make an ideal defense against ants, those omnipresent pests of the insect world. Ants, even those most aggressive fire ants, cannot pierce or puncture any part of the body, and those ants that clamp onto legs are easily brushed off with other legs. Meanwhile, the velvet ant hotfoots it out of there.

When an animal is so well protected and able to punish would-be assailants, all kinds of opportunities to prevent attack are evolutionarily possible. Why risk the slobber or damage of being in somebody's mouth if that can be avoided? Aposematic warning-signal systems go a long way toward achieving this. Once a predator learns an animal is unpleasant, it tends to avoid trying to eat that type of animal again. What better way to advertise your nastiness than color? Birds, lizards, amphibians, and most arthropods see color. Red and black are universal warning colors that signal to both experienced and naturally wary naive predators to leave them alone. The cow killer's flaming red and black beacons from grass, soil, or sand the advertisement "I am here—think twice before making a mistake." No matter whether the predator happens to be color blind, as many mammals are, the cow killer's pure red appears stark white in a black-and-white visual environment. The message, like the black-and-white message of a skunk, gets through.

Not all predators are visually oriented and may respond to warnings only after initial contact with a velvet ant is made. Velvet ants emit a warning signal in the form of a rasping stridulation against these acoustically or tactilely oriented predators. This signal acts much as the rattling of a rattlesnake's tail warns intruders. Both snake and velvet ant sounds are produced over an enormous range of sound frequencies, thus insuring the widest range of predators can hear them.[7] Mammals and birds are particularly sensitive to sounds. Most spider and insect predators either lack or have poor hearing; nevertheless, velvet ant stridulatory sound can be highly effective defenses against these arthropods. Hunting spiders grab or pounce on prey, simultaneously trying to puncture them with their fangs. Fangs are hard and inflexible. Velvet ants are hard and inflexible. The result is like applying a small jackhammer to one's teeth. The hard fangs, or mandibles, are released and the "vibrating rock" is rejected. Whether or not this vibratory defense works on its own with birds, lizards, or mammals is unclear.

Some predators, especially mammals, are highly attuned to odor as a clue to prey and edibility. Reptiles also possess acute senses of smell and taste, the contact analog of smell. When an insectivorous mammal or lizard grabs a velvet ant, the velvet ant releases its warning odor (along with stridulating and engaging the sting). After the predator experiences the sting, it associates the *eau de velvet ant* with the unpleasant sting and thereafter learns to associate the odor with the bad experience. The same probably applies to lizards, which usually lick prey before eating. A simple tongue flick to a velvet ant is usually enough for any lizard.

The warning odor may serve the dual roles of warning and chemical defense. Many species of velvet ants analyzed, plus those given the nose test in the field, produce the same blend of two major compounds, 4-methyl-3-heptanone and 4,6-dimethyl-3-nonanone, plus a few minor constituents.[8] The first compound is a well-known alarm pheromone/chemical defense in a wide variety of ants and even in daddy longlegs arachnids, those long-legged creatures looking like tan pills on stilts that cluster in cool, dark, damp areas. This sets the stage for multiple mimicry in which a diversity of velvet ants and other creatures all use the same chemical signals to warn predators that they are inedible. The compounds likely are also direct chemical defenses that taste somewhat like turpentine when eaten.

Do these velvet ant defenses actually work? Naturalists recognized years ago that velvet ants were generally left alone and not attacked. In 1921, the British naturalist Geoffrey Hale Carpenter, working in Uganda, tested a wide assortment of insects for palatability to a gray vervet monkey. The monkey was presented with an insect on the ground or in a box and his behavior observed. Insects expected to be palatable were generally eaten by the monkey. When a velvet ant was offered, he "pounced on it and hurriedly rubbed it on the ground in the manner previously described, eventually seizing it and crushing it up very quickly. I think his lips and one hand got stung. Another, smaller Mutillid [velvet ant] was then put down, but M[onkey] would not have anything to do with it." A month later he offered another velvet ant, and the monkey "with great eagerness seized it out of the box and bundled it into his mouth, getting his hands and lips stung. He shook his head and ran about, and the Mutillid fell out of his mouth, but he picked it up and ate it greedily, though for a few minutes afterwards he ran about shaking his head and wiping his mouth with his paws. M. must have been very hungry."[9] To test more species of predators, cow killers were extensively tested with many likely potential predators, including fire ants, harvester ants, a Chinese praying mantis, three species of wolf spiders, two species of tarantulas, four species of lizards, a bird, and gerbils (desert rodents from Asia).[7] Most predators attacked the cow killer; those that did not attack carefully eyed it before making no attempt to attack. Only 1 of 13 tarantulas and 1 of 8 gerbils ate a cow killer. The gerbils provided an interesting example of behavioral personalities of individual predators. None of the gerbils had seen a velvet ant before. Four of the eight were frightened by the cow killer, two gerbils attacked once, were stung and did not attack again, and two attacked twice with one abandoning attacks thereafter, and one succeeding in eating the cow killer. The behavior of the successful gerbil was particularly telling. That gerbil grabbed the cow killer, quickly rotated it in its paws while making quick bites into the spinning cow killer. It finally punched through the hard shell, inactivating the wasp, and then consumed it. This behavior was specific to the cow killer; other insects, for example, mealworm larvae, were simply held like a sausage and eaten from head down.

Spinning velvet ants brings to mind the University of California at Davis entomologist Richard Bohart. Bohart, a sturdy, intrepid student of wasps of all types, was a mentor to many young entomologists. The traditional wisdom was that velvet ants *could not* be picked up without being stung. Instead, one had to scoop them into a vial or jar. Dick, either too lazy or simply too tough, would casually, albeit quickly, roll velvet ants between his fingers and drop them into his collecting jar. Apparently, he applied the wisdom of gerbils and other insectivorous small mammals to his own collecting. We don't know if he was ever stung; perhaps if so, he kept stoic so as not to tarnish his reputation.

E. O. Wilson, the famous Harvard biologist, had a different experience with velvet ants than Richard Bohart. Ed was quite a bit younger than Dick at the time of his encounter with a velvet ant, an experience that may have helped fix him on a path of biology. "I could have been as young as three years old, and all I remember is that I was in this garden in somebody's backyard and I recall vividly this velvet ant—this mutillid running along—and my grabbing it and, of course, they have a horrific sting and it was so painful that to this day I remember the appearance of the garden, I remember the wasp, I remember how I felt. I don't remember anything else."[10]

A final way to measure velvet ant defenses is to examine the diets of highly insectivorous predators in the wild. The broad-headed skink, *Eumeces laticeps*, is a large powerful lizard that easily crushes large insect prey. Cow killers are abundant in its habitat yet were never found in the stomachs of skinks, despite their readily feeding on other noxious prey, including blister beetles, woolly bear caterpillars, ants, and stinging paper wasps. Twenty-three skinks were presented with cow killers. Eight skinks never attacked, nine attacked one to three times, and six attacked more than three times. Ten skinks were clearly stung, eight of which thereafter would have nothing to do with cow killers. Only two skinks killed cow killers, and just one ate the entire wasp. That skink attacked 23 times over 9 minutes and finally ate the intact cow killer because it was unable to crush its abdomen.[11] Meals of velvet ants are hard won.

The examples with cow killers and other velvet ants indicate something special about the stings and venom of velvet ants. What makes them so different from solitary cicada killers, mud daubers, iridescent cockroach hunters, or water-walking wasps? As so often in biology, much remains to be explored. What is known is that velvet ant stings hurt a whole lot more than stings of those other solitary wasps. Velvet ant venom is not particularly toxic to mammals, with a lethality 25 times less than honey bee venom, and a whopping 200 times less than an average species of harvester ants.[12] In its ability to destroy red blood cells, velvet ant venom is less active than the venoms of paper wasps and harvester ants by factors of 200 and 120. Clearly, cow killer venom is not special in ability to do damage. It also has low levels of the enzymes phospholipase and hyaluronidase but reasonably high levels of esterase.[13] What is special is its ability to produce pain. How, we do not know. I learned firsthand a year or so ago just how painful a velvet ant sting can be. I was innocently asleep in my bed one night when something tickled my upper leg. Reflexively, reaching down, I felt something hard, then bang, it hit me. Lights went on and an investigation revealed the pebble to be a small nocturnal velvet ant female who had objected to my rubbing her. The pain was sharp but had a rashy flavor to it—something commanding an urge to rub, but when rubbed, hurt more. No obvious redness or swelling other than that caused by the rubbing was apparent. In less than 5 minutes, the pain receded, allowing the possibility of sleep. She yielded a big punch for her size: a 1.5 on the pain scale.

If Dick Bohart could pick up velvet ants, why not me? After all, they are fast and even little ones are difficult to suck into aspirators, a sandy operation requiring time and much sand spitting. Large velvet ants are even more difficult to catch. These speedy fur balls will not fit into aspirators, are nearly impossible to catch with forceps, and when scooped into a jar or large vial are often accompanied by large quantities of sand and debris, something that requires removal later. Maybe Bohart had a point. So I began casually, the term being used loosely, picking them up and putting them in empty, cleaned peanut butter jars. Actually, the operation was more like holding the jar a few inches from the zigzagging, fleeing velvet ant, frantically pinching sand and velvet ant and raising the combo slightly higher than the jar before throwing it in the direction of the open jar mouth. Usually, this worked, and I merrily continued catching more. But one day while trying to get a striking black and orange *Dasymutilla klugii*, she nailed me. Although the sting was superficial, penetrating the skin for at most a few milliseconds, it produced an immediate sharp, rashy pain that once again produced this overwhelming urge to rub the site. The pain was mostly gone in 2–3 minutes, and entirely gone in 10 minutes, but the rashy feeling returned for days afterward when touched. Pain rating of 2 on the scale. Two months later, and not having learned my lesson, I was picking up an attractive glorious velvet ant, *Dasymutilla gloriosa*, a species clad in very long, fluffy white hair, and got the same thumb hit again. Sharp, intense, deep pain, again without causing redness or swelling, and the familiar rashy feeling and urge to rub ensued. Pain level also 2, but this time, by 6 hours a clearly defined area on my thumb had swelled into a tight pocket reminiscent of the medical term "compartment syndrome." This well-delineated tight area remained for three days and then receded. Out of sight, out of mind. Then, exactly two weeks later, all the skin in that area of tight swelling peeled off. Sometimes with stings strange, inexplicable reactions occur. This was one such reaction; was it an immunologically mediated Arthus reaction rather than a normal reaction? Who knows?

Plate 1. (*top*) A sweat bee, *Halictus* sp., on distant scorpionweed, *Phacelia distans*. Sweat bees are important pollinators that, when pinched, can deliver a mildly painful sting. Sting Pain Rating: 1 on the Pain Scale. Photo courtesy of Jillian Cowles. (*bottom*) Mud dauber wasps, *Sceliphron caementarium*, sting and paralyze spiders as food for their young housed in cells within "mud clod" nests adorning buildings and protected areas. These harmless wasps also frequent flowers and mud areas. They are nearly incapable of stinging. Sting Pain Rating: 1 on the Pain Scale. Photo courtesy of Margarethe Brummermann, http://arizonabeetlesbugs birdsandmore.blogspot.com/

Plate 2. (*top*) *Pogonomyrmex* harvester ants, historically demonized as destroyers of grazing lands, often facilitate plant growth and diversity around their colonies by enriching the soil with their wastes, clearing competing grasses, and discarding to perimeter refuse areas seeds that germinate, grow, and flower in the rich ant-formed microenvironment. Photo by the author. (*bottom*) Mating scrum of Pacific cicada killer wasps, *Sphecius convallis*. The male (*right*) is mating with a female (*left*) that is also supporting a smaller male, the last of many to give up in the competition. These impressive, harmless wasps do not sting unless handled roughly and all-too-frequently are mistaken as huge yellowjackets. Sting Pain Rating: 1 on the Pain Scale. Photo courtesy of Chuck Holliday.

Plate 3. A male tiphiid wasp jabs its pseudosting into the author's finger. Males of stinging insects lack a true stinger, but some species are able to jab their sharp pseudostings into captors, thereby startling the captor into releasing the harmless male. Sting Pain Rating: 0 on the Pain Scale. Photo by the author.

Plate 4. Justin Schmidt in 1975, excavating a Florida harvester ant, *Pogonomyrmex badius*, colony in Amite, Louisiana. Sting Pain Rating: 3 on the Pain Scale. Photo by Debbie Schmidt.

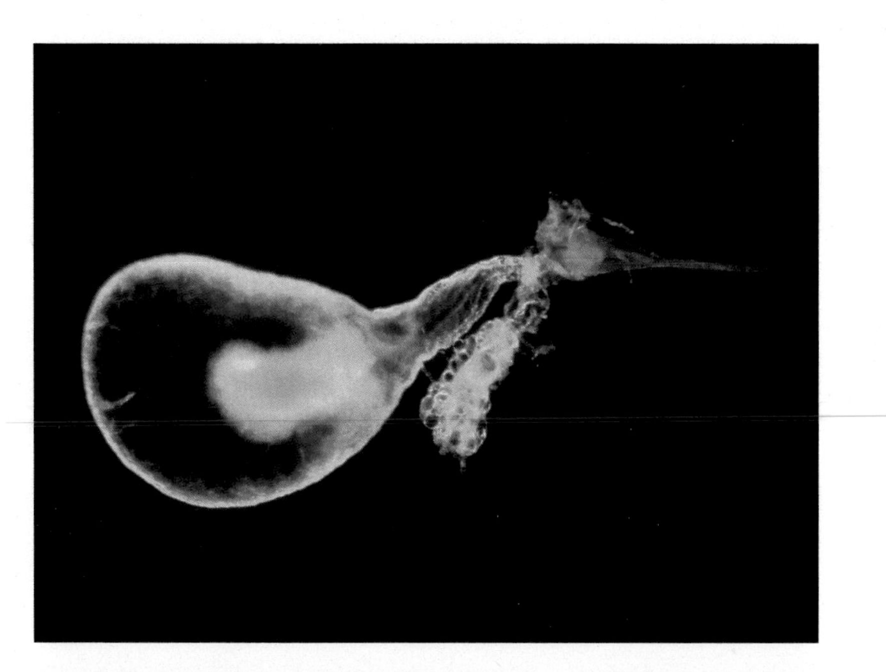

Plate 5. (*top*) Fire ant, *Solenopsis invicta*, sting apparatus illustrating the thin, sharp, needle-like stinger attached to a large venom-filled reservoir and a smaller frothy-looking Dufour's gland. The piercing stinger and enormous venom reservoir constitute an ideal system for injecting poison into assailants. Sting Pain Rating: 1 on the Pain Scale. Photo by the author. (*bottom*) Honey bee leaving its stinger in a victim's arm. Photo by the author.

Plate 6. (*top*) Tarantula hawk, *Pepsis chrysothemis,* collecting nectar from desert milkweed, *Asclepias subulata.* These brilliantly colored, conspicuous solitary wasps are unaggressive but not to be held. Sting Pain Rating: 4 on the Pain Scale. Photo courtesy of Jillian Cowles. (*bottom*) Tarantula hawk and its tarantula prey. In these fierce battles, the tarantula essentially always loses. Nic Perkins photo courtesy US National Park Service.

Plate 7. (*top*) Female velvet ants are wingless solitary wasps that are often colorful and seen in open areas during the summer. They range in size from small, as is this 6-mm *Dasymutilla asteria*, to huge nearly 25-mm "cow killers." Sting Pain Ratings: 1–3 on the Pain Scale (depending on size of velvet ant). Photo courtesy of Jillian Cowles. (*bottom*) Bullet ant, *Paraponera clavata*, a species universally feared and respected wherever it occurs. These ants are sometimes a part of puberty rites in local populations in Amazonas. Sting Pain Rating: 4 on the Pain Scale. Photo courtesy of Graham Wise.

Plate 8. Sample collection as part of a study in Costa Rica of the genetics of Africanized honey bees. These bees were intentionally provoked, something not recommended for the inexperienced. Photo of the author by Hayward Spangler.

10

BULLET ANTS

I can only liken the pain to that of a hundred thousand nettle stings. —Richard Spruce, *Notes of a Botanist on the Amazon and Andes,* 1908

When I picked her [a Paraponera] up, she stung me and immediately it felt like someone had smashed my thumb with a hammer. —Marlin Rice, 2014

Bala, tucandéra, conga, chacha, cumanagata, munuri, siámña, yolosa, veinte cuatro hora hormiga, bullet ant. These are some of the common names for *Paraponera clavata,* the world's most painfully stinging bee, wasp, or ant. This impressive giant ant, with its stocky, black body and impressive jaws and stings, is known and given a local name by people wherever it occurs. But don't let this ant's primitive appearance deceive you into thinking it is just a slow, dim-witted brute. It is a lithe, arboreal acrobat, all too ready to demonstrate its agility as it clings and stings. *Paraponera* knows no fakery, it's the real thing. *Paraponera* is the insect star in stories worthy of telling to one's grandchildren and in the 2015 movie *Ant-Man*. If stung, you might not think you will live to see grandchildren, but, rest assured, no one has ever died from bullet ant stings.

The bullet ant, *Paraponera,* inhabits moist forests stretching along the Atlantic Ocean side of the continental divide, from Nicaragua in Central America to Brazil in South America. On my first visit to the

Costa Rican tropical research station at La Selva, I was greeted by a conspicuous sign warning to be exceedingly cautious of the stinging bullet ants. This struck me as odd, given that the station nestled in a rainforest was home to dangerous and potentially lethal fer-de-lance snakes. No signs were planted in the grassy areas around the station about fer-de-lances. My quest at the station was to study the marvelous bullet ants and their defenses and venoms. As an entomologist, I had a healthy respect and admiration for these ants and immediately set out to find them. Because bullet ants are active both day and night, I headed out at night equipped with headlamp, jars, and the usual insect net. An army ant column had been seen crossing the path earlier, so I made a side trip off to find their bivouac. Going through the undergrowth was tough. About five minutes in, I heard this "flop, flop," emanating from somewhere in the leaves in front of me. At first, nothing was visibly causing the sound. Then I saw it. A huge 2-meter-long fer-de-lance was rearing its head off the forest floor and flopping down into the dry leaves, hence the flopping sound. While elevated, the snake's mouth was open. Fair enough, the snake was warning me in two ways not to step on it. Without the warnings, the snake was perfectly camouflaged and invisible among the leaves. After admiring it and taking some pictures, I decided the only safe way to proceed was to put the snake in my insect net (a fine herpetological tool) and hold it at arm's length in front of me. That way, I would always know where it was and couldn't step on it. It was heavy. A 10-pound snake held 6 feet away quickly becomes a tiresome challenge. The snake interaction caused me to lose track of my course, and the army ants were nowhere in sight. Unless I wanted to spend the night alone and lost in the forest with a big snake, it was time for plan B: the snake was unceremoniously tossed downhill, and I headed uphill to the path. That was it for the night. I asked the local herpetological expert at the station about the fer-de-lance. He asked, "Does it have keels [central ridges on each scale] on its scales?" "Are you kidding? Get that close to see keels?" He figured it was not a fer-de-lance at all, as they have no keels, rather a bushmaster. Bushmasters are the largest venomous snakes in the New World, attaining a length up to 3.5 meters, and are the most deadly Costa Rican snake. At that time, six of the seven recorded bitten people had died. He surmised I had run into a "small" bushmaster. Sure enough, when the film was processed, the scales had keels. And the station warns about bullet ants!

Bullet ants, though not lethal, are impressive. They left indelible images in the minds of early naturalists and their readers. One of the earliest writers was the botanist Richard Spruce, who wrote of his experience on August 15, 1853, in Amazonas:

> Yesterday I had the pleasure for the first time of experiencing the sting of the large black ant called tucandéra in Lingoa Geral. . . . I did not notice that a string of angry tucandéras poured out of the opening I had made. I was speedily made aware of it by a prick in the thigh, which I supposed to be caused by a snake, until springing up I saw that my feet and legs were being covered by the dreaded tucandéra. There was nothing but flight for it . . . but not before I had been dreadfully stung about the feet. . . . I was in agonies, and had much to do to keep from throwing myself on the ground and rolling about as I had seen the Indians do when suffering from the stings of this ant. . . . I can only liken the pain to that of a hundred thousand nettle stings. My feet and sometimes my hands trembled as though I had the palsy, and for some time the perspiration ran down my face from the pain. With difficulty I repressed some inclination to vomit. . . . After the pain had become more bearable [3 hours], it returned at 9 o'clock and at midnight, when I stepped out of my hammock on my left foot, and each time caused me a hour of acute suffering. . . . It is curious that nothing was visible externally more than would be caused by the sting of an ordinary nettle. . . . I came worse out of this encounter than any other in which I have been engaged since entering South America. Many times have I been stung by ants and wasps, but never so badly.[1]

Spruce was not alone in noticing the sting of bullet ants, so named because victims sometimes likened the pain to being hit by a bullet. Algot Lange wrote in1915 of his trip on the Javary River tributary of the Amazon River during which he was stung on his leg by a bullet ant: "The pain almost drove me out of my senses for fully twenty-four hours, and the inflammation abated only the third day after the bite [*sic*]. When the Brazilians declare that four *tucandeiras* will kill a man I believe it; while perhaps he will not die from the actual poisoning, he might from the agony associated with such bites [*sic*]."[2] Hamilton Rice, a medical doctor, writing a year earlier of his explorations on the northwest Amazon, offered these observations: "The insects and pests of these regions make existence a continual torment and seriously interfere with the power of work. The worst and most dreaded of the ants is the tucandéra or conga, the bite [*sic*] of which causes the most excruciating agony lasting for several hours, sometimes attended by vomiting and hyperexia [hyperthermia]."[3] Apparently, bullet ants made a greater impact on people than the diseases of yellow fever, malaria, or river blindness.

Nearly four decades after Rice, Harry Allard, a Washington, DC, botanist, writes of attempting to pick up *Paraponera* (misidentified as *Dinoponera*, an even larger ant) using several folds of a handkerchief. He describes the ant as "a handsome shining black insect an inch or more in length and fears no one" and continues describing the sting to the end of his index finger: "the pain was soon excruciating and lasted until well into the night. So severe was the pain that at times my hand trembled. The next day there were redness and swelling, but no other local symptoms." Some weeks later he was stung twice on the ankle "and in a short time I was in the throes of an agony of burning pain— a pain such as I have never experienced before, nor ever care to repeat. . . . I could not keep my foot quiet for any length of time." Allard goes on to describe a sting to his 3-year-old grandson who "with a child's curiosity" had picked one up, the pet dog's sting to a paw, and the mortal fear of two white-faced monkeys.[4] Fast-forward 60-some years to Terry Erwin, the Smithsonian Institution biologist who invented canopy fogging to survey insect biodiversity in tropical forests and has spent more time in bullet ant territory than anyone I know: "I've seen snakes and all that kind of stuff and been stung by *Paraponera*. It's a *real* shock when you get stung and you know *immediately* [his emphasis] what it was. . . . So I grabbed that thing and pulled it out. . . . I was just squeezing and squeezing, and then it dropped out, but they are so hard I didn't kill it and it was crawling away. That lasted about half an hour, and by day two, oh, then after the fire, it goes to feeling like a dull toothache and the toothache kind of goes for a couple of days."[5] Enough belaboring the point, I think we can appreciate that bullet ants are no ordinary ants.

What makes these ants so unusual? An adventure into ant taxonomy and bullet ant natural history provides clues. The 15,000 or so described ant species currently fall into 16 subfamilies. Until recently, only 9 subfamilies were commonly recognized, including the Ponerinae, the third-largest subfamily. This subfamily was a miscellaneous trash can of species, lumped together largely based on their heavy body design, simple colony structure, with simple behaviors, and other "primitive" traits. *Paraponera* was lumped in this subfamily, albeit a somewhat unusual species given its nasty sting, but otherwise was "just another ponerine ant." Hence, its behavior and venom were compared to those of other ponerine ants. Why, then, was it so different? In the past decade, genetic analyses of ants have revealed that *Paraponera* was not in the subfamily Ponerinae after all. And its previously presumed closest sister group, the genus *Ectatomma*, was taxonomically even more distant, more closely related to the formic acid–spraying carpenter ants of one's backyard than to the ponerines. What this taxonomic diversion tells us is that the bullet ant, now in its own subfamily of one species, is truly a unique ant whose lineage separated from other ants around a hundred million years ago.[6] Thus, *Paraponera* should not be expected to be especially similar to other ants, no matter how similar these other ants might superficially look, or whatever their taxonomy.

The biology of bullet ants differs from that of other ants. They live mainly in nests in the ground near the base of trees. Although their nests are in the ground, they do not forage on the ground around the nest entrance, preferring to climb up the trees into the forest canopy to forage. Sometimes they will climb into the canopy, only to return down a different tree or some of the abundant vine-like lianas to the surface, where they forage at distances as great as 60 meters from their nest entrance. This foraging behavior seems to be a way to prevent leaving a trail or other clues for potential assailants or competitors to locate the colony. Sometimes colonies will be located aboveground in the forest trees; these nests are usually located in sizable quantities of debris and humic material that has a consistency similar to soil that accumulate in the crotches of large trees. In Costa Rica, newly mated queens tend to locate their colonies near one particular species of tree, *Pentaclethra macroloba*, apparently based on chemical odor.[7,8] This nesting behavior apparently reflects local conditions, as they are opportunistic and flexible ants. This nesting flexibility was apparent on Barro Colorado Island in neighboring Panama, where colonies were associated with 76 species of trees, shrubs, palms, and lianas, none of which were *Pentaclethra*, a species not present at the site.[9]

Bullet ants are no velociraptors of the ant world; they are mainly vegetarians, consuming sugary solutions of sap, fruit juices, and other unknown sources present in the forest canopy. Unfortunately, sugars do not provide the protein necessary for growth of bullet ant larvae or egg production by the queen. To meet these protein needs, bullet ants also become predators of an assortment of insects, spiders, and other invertebrates. Even hard, spiny, biting leaf cutter ants—those big-headed orange ants seen in nature programs moving in long columns, carrying green leaf flags as they march home—are taken.[10] Foragers, typically larger individuals from among a continuous worker size range of about 15 to 22 millimeters, are, nevertheless, choosy in what prey they will accept. Many chemically protected caterpillars or other noxious prey, including the strawberry (colored, not tasting!) poison-dart frog, *Oophaga pumilio*, are rejected. The poison-dart frog is rejected not because it is a frog but because of its taste; similar-sized, cryptically colored frogs in the genus *Eleutherodactylus* are accepted.[11]

Bullet ants, though large, are not primitive or the dumb brutes of the ant world. They live in large colonies of up to 2,500 individuals and are as capable as honey bees in learning timing of food rewards and exhibit learning based on experience and orientation cues.[12] When a rich source of food is located, bullet ants can recruit nestmates to the source via a chemical trail pheromone they lay down by rubbing their abdomen on the surface as they return.[7,13,14] They can even evaluate the benefits and costs of recruitment based on the concentration of sugar solution and the travel distance.[15]

As idyllic as bullet ant life appears, all is not always well. Like our own species, their worst enemies might be themselves. Sometimes huge fights between colonies erupt with a dozen or so enemy pairs engaged in often mortal combat.[16] Intercolony conflict results in overdispersed nests in the environment, that is, not randomly dispersed as in the BB holes in a shotgun target. Instead, they more uniformly separated from one another. Colonies closer together than 20 meters have significantly higher mortality rates than those farther apart. Mean colony life expectancy is only 2.5 years, with intercolony aggression a major factor contributing to the short life expectancy.[17]

Other than neighboring colonies, bullet ants have few natural predators. I have observed that other ants, even army ants, do not seem to bother them. The one record of vertebrate predation on bullet ants that I could find was reported in 1943 by Albert Barden. Barden inventoried the stomach contents of a large number of basilisk lizards, not the huge basilisk snake of Harry Potter fame, but midsized lizards commonly known in Central America as Jesus Christ lizards for their ability to run on water. Among the 1,141 separate food items found in basilisks were a few bullet ants.[18] Whether these were active foragers or were injured combatants fallen from trees, we will never know. In any case, vertebrate predation in nature of bullet ants is rare at best.

Meet the Toad Project. While I and two colleagues were in the Guanacaste dry forest of Costa Rica, we took a break from studying "killer" honey bees (Africanized bees) and ventured over the mountain spine to the Atlantic rainforest. There, some bullet ant workers were collected and brought back to Guanacaste. Around the dinner table at La Pacifica where we were staying was an abundance of giant cane toads, *Bufo marinus*. Toads are among the most indiscriminate predators and are deterred by very little. If it moves, it is eaten. Hugh Cott reported his tests on the palatability of honey bees to common English toads, *Bufo bufo*, in 1936. He determined that toads readily eat the first presented bee and some learn after one or more stings that bees are a bit too spicy for them. Other toads repetitively ate bees despite receiving up to five stings before avoiding more bees. His study showed that toads are hardy, readily endure stinging punishments, and are sometimes slow learners, but eventually within seven days all learned that bees are not preferred food.[19] Given that toads seemed to be the toughest predators around, and were readily available around our table, we decided to test the palatability of bullet ants to them. We chose a random, good-sized toad and tossed it a bullet ant. Down the hatch went the ant. The toad responded by "hiccupping" body jerks, eyes bulging in and out, and mouth gaping. Apparently, the toad was stung. Would the toad learn? No. Down the hatch went the second bullet ant. Same toad reaction. Did it learn after two ants? No. Down went a third, with the same reaction. The toad ate nine bullet ants in a row, each time reacting to the stings. At this point, we had run out of bullet ants, so we tossed a nonstinging insect to the toad to see whether the reaction would be the same as for a bullet ant. Down it went without a hint of discomfort. Toads are tough, and apparently might possibly be predators of bullet ants, though something not reported from the field. This table-side test shows how extreme a predator must be to tackle a bullet ant.

Perhaps the greatest impediments to the life of bullet ants are not predators but tiny parasitoid flies. As mosquitoes torment human lives, flies, *Apocephalus paraponerae*, torment bullet ants. These flies are roughly the size of common vinegar flies that love our overripe bananas and are cherished in genetic research labs. Unlike vinegar flies, they deposit their eggs on injured bullet ants. When one of these flies hovers near a bullet ant nest entrance, in the words of one writer, "more than 10 ants came rapidly out of the nest and went berserk trying to catch it."[20] Within minutes after an injured ant is presented, flies come from seemingly nowhere to the ant. Both male and female flies arrive. Females oviposit on the ant, the maggots consume the ant, and up to 20 flies result.[21] We asked, How do the flies locate the injured ants so quickly? A promising lead came from the odor released by injured ants. Bullet ants contain in their mandibular glands a ketone, 4-methyl-3-heptanone, and its corresponding alcohol. An injured ant releases these odors. To test the possibility that these chemicals were attractants, we added them to highly refined olive oil to make a system for slow chemical release. Sure enough, flies were attracted to the baits.[22] That these flies have a highly specific and effective means of discovering and exploiting their food source indicates injured ants are not scarce and that battles between individual bullet ants are a major source of mortality.

The stinging power of bullet ants did not go unnoticed by local people. Various indigenous peoples in northern Amazonas traditionally used and some continue to use the ants in ceremonial rites of manhood. The Ararandeuara used a ceremonial braided fiber cylinder about 60 cm long and 20 cm in diameter, closed at each end with a drawstring. The cylindrical muff was filled with bullet ants, the desirous youth then placed his hand in the mitt, and the drawstrings were snugged around his forearm. If he could keep his hand inside the mitt for a length of time, withstanding the pain, he was proclaimed a man fit to marry and the celebrations continued.[2] This procedure adds perspective to the old adage "no pain, no gain."

Several variants of puberty rites have been reported among the Amazon peoples. In Suriname, woven mats are fashioned to hold bullet ants by the constriction between the ants' thorax and abdomen, thereby preventing ant escape. The mats are then charged with bullet ants having their business ends all facing one side. The mat is placed on the candidate boy's abdomen, gluteus maximus, thighs, and so forth, where they readily sting. After withstanding this treatment "like a man," the boy is fed an herbal concoction and rests in a hammock, while the rest of the tribe has a long celebration of his manhood.[23] Another version of the ceremony was recorded on video in the PBS *Nature* series titled "Gremlins: Faces in the Forest," produced in 1997. The main emphasis was discovering and recording the tiny and reclusive marmoset and tamarin monkeys of the Amazon forest. In the process of their discoveries, they recorded "as filler" the use of bullet ants in a manhood ritual. In that version, the boy was first painted black before the ants were applied. Several YouTube clips have appeared subsequent to that 1997 "Gremlins" recording. These included a ceremony featuring a Satery boy during initiation rites (http://www.youtube.com/watch?v=XwvIFO9srUw). His dedication and mental strength make the ceremony appear perfunctory and the stings only a minor pain. Another view narrated in Portuguese (http://www.youtube.com/watch?v=gjna7MnPKrI) reveals a greater impact of the pain. Hamish and Andy provide an ultimate contrast in human response to the sting pain; in this case, by a person not so dedicated to passing an initiation test (http://www.youtube.com/watch?v=it0V7xv9qu0&list=RDit0V7xv9qu0&index=1). Several hundred thousand viewings of these ceremonies suggest a great underlying appeal and popularity of bullet ants to people, no matter where they live.

The Ka'apor of Amazonas use another large ant, the termite predator *Neoponera commutata,* in puberty rites for girls. The stings of these ants are painful, but not nearly as painful as those of bullet ants; hence, their choice for girls. William Balée of Tulane University described these rites in detail, along with several others involving bullet ants and other ants.[24]

The Amazon people appreciated the toxicity of bullet ants. Some tribes in the far upriver reaches of the Amazon mixed bullet ant venom with other toxic ingredients to produce the arrow poison curare, locally known as woorali, which is deadly if introduced under the skin, but harmless when ingested.[2] My suspicion is that the painless curare alkaloids were the true paralytic/lethal components and the bullet ant venom contributed the painful part of the poison. What tribal peoples knew for certain was the painfulness of bullet ant stings and venom.

The culprit that ties together all the stories about bullet ants is the venom. Bullet ant venom is as unique and unusual as the ants themselves. The venom is highly lethal to mammals, having a lethality of 1.4 mg venom per kilogram body weight, and is produced in the prodigious quantity of 250 micrograms per average ant.[25] The two combine to yield a projected capacity for one sting to kill a mammal of 180 grams, about the size of a young female Norway rat. This killing power is over three times that of a honey bee and nearly eight times that of a baldfaced hornet. In contrast to the venom's killing power is its amazingly low ability to destroy cell membranes and tissues. Among the 10 ants tested, bullet ant venom came in dead last in ability to destroy red blood cells, a standard assay for tissue damage. Its activity was so low that it was 48-fold lower than harvester ant venom and a whopping 1,200 times less active than the Brazilian paper wasp, *Polistes infuscatus*.[26] This low membrane and tissue damage potential explains the typical human sting reaction of little swelling or redness and trivial signs of the sting after the pain recedes.

Two questions spring to mind: (1) What makes the venom so painful? (2) What makes the venom so lethal? The venom contains small amounts of kinins, much like those abundant in social wasps, but these are minor factors in the overall activity and do not produce redness or swelling, which occurs with wasp stings. The more interesting factor is a 25 amino acid peptide called poneratoxin. Poneratoxin is four times more lethal than whole venom and accounts for most of the venom's lethality. This acidic peptide is also highly active at concentrations as low as 25 μg in a liter volume, causing long-lasting contraction of smooth muscles in the body, undulating changes and bursts of transmitter release in nerves and muscles, blockage of cockroach nerve signal transmission, and impeding sodium channels of skeletal muscles.[27] These activities explain most, if not all, activities observed in the field.

As a final check on the activity of poneratoxin, I injected a tiny amount of synthetic poneratoxin provided by my collaborator Steve Johnson under my forearm skin to make a small bleb about a tenth the size of the bleb produced in a TB test. The reaction and pain were identical to that of a real sting but, fortunately, less severe than a true sting because only a small amount of venom was injected. (I wanted answers, but not serious pain; that's why the dose was small.) The dose was too small to cause muscle trembling, something predicted for poneratoxin, though it did induce the urge to shake the arm. This forearm test suggests that most, perhaps all, of the sting pain and reaction is caused by poneratoxin, though other causative factors could also be present. No other ant, or other venomous animal, contains a peptide similar to poneratoxin. It is a truly unique toxin from a truly unique ant.

The superlative nature of this ant and its venom brings to mind the question, Why? Why does this species need such a potent venom? Why don't other ants (or other stinging insects) have similar venoms? These questions cannot be answered directly, but we can get hints about the forces operating on the ant that selected for its venom. These forces, primarily from large vertebrate predators, operate to a lesser degree on many other stinging social insects. Why only bullet ants evolved poneratoxin but not other ants, wasps, or bees is part taxonomic lineage and part randomness. Once a trait evolves in a lineage, descendant species of that lineage can easily maintain or genetically modify that trait. Neither bees, wasps, nor other ants had anything similar to poneratoxin in their lineage, so evolution of a poneratoxin-like molecule would have to be de novo, a much more difficult evolutionary process. The bullet ant lineage separated from those of other ants around 100 million years ago. This provided bullet ants a hundred million years alone to evolve poneratoxin, something they obviously did, perhaps by random mutation.

Back to the forces that would select for poneratoxin. In rainforests, most vertebrate predators, whether mammals, birds, lizards, or frogs, live in the forest canopy. Relatively few live on the dark forest floor or on dim low understory plants. Part of the reason vertebrates are active in the canopy is that most leaves, flowers, fruits, and insects are present in the canopy. To get these resources, a species must go to, or be in, the canopy. However, for an insect, the canopy is a dangerous place where you are almost as likely to become dinner as find dinner. The insects in the canopy are generally hidden in retreats much of the time or are camouflaged, cryptic, behaviorally secretive, temporally ephemeral, or otherwise out of view. The alternative is to be bright, flashy, and very nasty. Bullet ants nest in the relatively safe ground environment but must forage in the canopy for food. They are big, conspicuous, and long-lived, all factors working against survival among hordes of hungry birds, monkeys, lizards, and amphibians. What bird or monkey would not like a nice big, juicy insect for a snack? Bullet ants cannot jump or fly away to escape, and they cannot easily hide, so they must face the predators. This they achieve better than any other stinging insect through their venom. If a predator is unwary enough to grab a bullet ant, it likely will remember the experience and not repeat it. To aid in warning predators to look elsewhere for food, bullet ants have several aposematic signals. First, they are shiny and blackish, a common indication of inedibility. Second, they loudly stridulate, producing a sound warning to anyone nearby that they are present and not to be messed with. Third, they produce 4-methyl-3-heptanone and other chemicals warning want-to-be predators to be wary and avoid them. In addition to these three warnings are undoubtedly more behavioral traits that signal that they are bullet ants. Some predators are tough, as the toads mentioned earlier; some predators are smart, as are monkeys, and learn tricks to overcome prey. All these predators are in the canopy. To forage day and night in the canopy, bullet ants need serious protection, and their sting is the best there is.

I am often asked how I know bullet ants are the most painful of any stinging insect. Of course, this can never be answered with 100 percent certainty, as thousands of stinging insects have been described, with more awaiting discovery. Neither I, nor anyone else, have been stung by all of them. I have searched six continents (skipping Antarctica) over 40 years for stinging insects and never found any whose stings come even close to the pain level and duration of bullet ant stings. This is not from lack of specific search. In South Africa, I tracked down the feared matabele ant (*Megaponera analis*), the giant stink ant (*Paltothyreus tarsata*), and others and found their stings mild compared to bullet ant stings. In Australia, the bull ants in the genus *Myrmecia* are believed to have horrific stings. Yet, when I and others were stung by bull ants, the pain did not equal that of a honey bee, much less that of a bullet ant. The famed bull-horn acacia ants (*Pseudomyrmex*) hurt, but again, much less than bullet ants. Tarantula hawk stings hurt as much for a couple of minutes as bullet ant stings, but then vanish, something all of us wish would happen with bullet ant stings. There are reports of extremely painful stinging ants in the Congo and wasps and ants in the western Amazon basin, but the low number of these specific reports and their infrequent or nonexistent confirmation suggest they, too, would be less painful than bullet ant stings. Reports of bullet ant stings almost universally expound the painfulness of the stings. I am confident that bullet ants are the holy grail of stinging insects and deliver the most painful sting of any stinging insect on Earth.

Have I ever intentionally stung myself with a bullet ant? Of course not. No need to. Bullet ants are all too ready to oblige. Mess with a colony, and you will likely be stung. Innocent in the rainforest? Watch out. A hand on a sapling, liana, or tree buttress is asking to be stung by an unnoticed bullet ant. Experienced people learn to look before leaning and, in general, not to grab or hold anything unless it is necessary. The slouch leaning on a fencepost or a tree trunk will quickly stand up straight.

My first experience with the famous tucandéra, as bullet ants are called in Brazil, occurred in the delightful city of Belém at the mouth of the Amazon River. I was with my professor, the colorful, talented Murray Blum, and colleague Bill Overal of the Emilio Goeldi Museum. We were in an older second-growth forest with a goal to collect as many ants and wasps, especially the stinging kind, as possible. We wanted them for comparative studies of pheromones and venoms. We had along an able-bodied assistant named Romero. Romero was the kind of guy you want on your team: big, strong, and afraid of nothing (or so I thought). Find a fire ant colony. No problem. Romero would grab handfuls of dirt with ants, stuff them in a plastic bag, and brush off any remaining ants. Find a social wasp nest in a bush. No problem. Romero would grab it and stuff it in another bag and then slap away those wasps chasing him. We found *Dinoponera,* those gentle giants of the ant world and the largest ants on earth, and allowed them to crawl on our hands and faces. We then found a bullet ant colony at the base of a sapling. Just what was wanted. They were too large to suck into my aspirator, so I was collecting them one by one with long, 12-inch forceps. That was difficult. Bullet ants are remarkably quick, strong, and agile. And they are sticky, clinging amazingly well to polished chrome forceps as they move ever closer to fingers. I managed to collect all the individuals around the entrance without getting stung, but as daylight was ending I still wanted a lot more before having to leave for dinner. Hurried digging with my pathetic trowel wasn't working. "Romero, I need help cutting these roots with your mattock. Where is Romero?" Turned out Romero (and Bill and Murray) were safely at a distance watching. "Romero, I need help!" Romero charged in, took a few whacks at the roots, and retreated. Time was getting short, light was fading, and ants were boiling out. No choice, forceps weren't working well. Grab an ant and with lightning speed toss it into the talcum-powdered, escape-proof jar and repeat. Only bullet ants are lightning fast, too. I can't remember exactly how many stings I got, four I think, but they were absolutely excruciatingly painful and debilitating. Enough ants—out of here! Bill knew a local churrascaria, one of the Brazilian restaurants that specialize in dozens of different types of meat, many stuck on long swords, that are brought to your table. As we drove to the restaurant my hand was throbbing, sending crescendos of pain, followed by easing a bit, only to be repeated with renewed ferocity. All the while the forearm was uncontrollably vibrating up and down. "Stop it, damn it!" No matter how hard I tried, I could not stop that hand and arm from trembling (the other arm was fine). When I touched the skin around the main sting, it felt numb. Even when I poked the area with a pencil point, I detected no sensation. If poked hard enough, I sensed a dull, deep visceral pain, but nothing else.

When we arrived at the churrascaria, my first request was ice. My second request was a beer. Ice did actually arrest most of the pain, and the beer provided a lift in spirits. Another beer later, it seemed time to stop the ice. Good food was on the table and eating with an iced hand was difficult. As the cold wore off, the pain resumed unabated. It seemed as if I had only delayed the clock, the pain was going to get its time no matter what. Ice back on. Dinner finished and time to relax and plan for the next day. The pain was still there. Bedtime. The pain was still there. Getting to sleep was difficult, but finally came around midnight in spite of the pain. The next morning the pain was finally gone, and if to add insult, little sign I had ever been stung was revealed. Another encounter with bullet ants shows just how effectively they can defend themselves. We were in Costa Rica excavating a bullet ant colony. This time there was no haste and neither of us was interested in being stung. Meticulous excavation yielded many ants and no stings; that is, until an ant fell from the liana above, bounced off my cheek, and landed on the ground. In the process of bouncing, it managed to sting my cheek. The sting was minor, given the little time to inject venom, but it was still a sting. Bullet ants are quick, something we humans are quick to learn through experience.

11

HONEY BEES AND HUMANS

AN EVOLUTIONARY SYMBIOSIS

Apis dorsata is the most ferocious stinging insect on earth.
—Roger Morse, "*Apis dorsata* in the Philippines," 1969

Baby toys, elephants, yellow rain, and *Apis dorsata*. What do they have in common? They all are connected to honey bees, or they *are* honey bees. Humanity has no greater a mixed relationship with any animal than with the honey bee. Honey bees are prominently featured in several major religions and are the state emblem and insect of Utah. Israel is the Land of Milk and Honey. Ah, honey. That gives us a clue. Honey bees produce honey and people love honey. Therefore, people love honey bees. But wait. Honey bees sting! That is the crux of our tormented relationship with honey bees. We love their honey, and we love them because they make honey. Yet, they sting, and we are afraid of their stings. The two combined make honey bees most fascinating insects and the second-most scientifically studied of all insects (only the fruit fly, *Drosophila*, that "white rat" of the insect world, has been the subject of more publications).

My own life with honey bees started at an early age—the exact age I cannot remember. I discovered that I had a talent for picking up honey bees from clover flowers without getting stung. Although

I somehow knew honey bees could sting, I do not recall ever being stung; however, my teacher had quite a different experience. I did not remember the incident on the playground in which I placed a honey bee on her arm, but my teacher and mother remembered it and regaled me with it many times during my school years. It turned out that my mother and the teacher were friends.

My first memory of a sting was not from a honey bee but from a bumble bee. Bumble bees, like honey bees, are beloved insects whose motifs adorn infant clothes and toys. One of the reasons they are beloved is that they, with their soft yellow and black rounded furry bodies, are cuter than honey bees. Who doesn't remember seeing bumble bees playfully visiting one flower in the yard and then the next? Bumble bees also tend not to be aggressive like honey bees. That is, unless one messes with their nest. Then, like any good parents, they tend to defend the home and their young. Mess with their nest is exactly what I did. I was five years old, and our backyard had a low pile of wood in a corner. I saw bumble bees entering and leaving from deep within the woodpile and decided to investigate where they were going. The memory of how I disturbed the bumble bees is lost, but I remember the consequences clearly. Bees came out and attacked. One bee managed to attach itself to the back of my neck and sting me. I screamed and ran toward the back door, swatting at the bee on my neck as I ran. Unlike honey bees that sting only once and lose their stinger in the process, bumble bees do not lose their stingers and can sting many times. That day I received five stings to the back of my neck by that one bee. After that, I never messed with the bumble bees in the yard again.

I don't remember my first honey bee sting. After the bumble bee sting, I returned mostly to climbing trees, playing in the nearby brook, exploring the local woods, and, a favorite, roaming through the meadows and abandoned old fields. I never really thought of honey bees. I was much much more interested in butterflies and other colorful insects that visited the flowers. My dad, a practical man of many interests and talents, took up beekeeping for fun. First, he had only a couple of colonies, then half a dozen or so, and finally about 40. Naturally, my brother, sister, and I enjoyed the honey and helping during honey-extracting time. First, my brother, who was two years older than I, got a hive of his own. Not to be outdone, I got my first two hives the next year. Those were good times. I was in the 4-H beekeeping club, earned a Boy Scout merit badge in beekeeping, and my sister was the Honey Princess (then called First Runner Up) in the National Honey Queen contest. My brother got progressively larger swellings after being stung, and after a sting to a hand caused swelling past his elbow (we chided him that he was Popeye, the Sailor Man), he gave up beekeeping. I must have been stung during this time, but I cannot remember. Perhaps joy overcomes pain.

Honey bees went into space with NASA—not just once, but twice, first in March 1982, then in April 1984. The second time, the bees built in that zero-gravity environment a normal honeycomb that appeared like combs built under normal gravity here on Earth. They also stored nectar and laid eggs in the comb, as normally expected. The fascinating biology of honey bees is what interested NASA in sending bees into space. Honey bees represent a pinnacle of social evolution within insects in which they live in large colonies of 15,000 to 30,000 individuals (with a range from 1,000 to about 60,000) comprised mostly of non-reproductive worker bees plus a few males, unflatteringly called "drones," and a single egg-laying queen. Colonies are perennial and reproduce by swarming, in which a portion of the workers and a queen leave the parental nest and form a new colony in a different location. Queens, as needed to replace an old queen, or to leave with swarms, are reared in special peanut-shaped cells and fed a special creamy white diet called royal jelly. Unlike most wasps that are carnivores, honey bees are strict vegetarians (vegans, if you wish) that feed on pollen and nectar from flowers, plus sugary liquids from other sources, including honeydew-producing insects.

A unique feature of honey bees is their use of hexagonal cells of honeycomb constructed only of the beeswax they secrete from wax glands on their abdomens. The elegant design of these combs with their mathematical simplicity and economy of material has, over the years, attracted the attention of scientists, including Aristotle and Charles Darwin. Exactly how bees measure and coordinate the nearly perfect geometrical construction of the comb has been a source of human intrigue, including by NASA. NASA's question was, "Can bees make a perfectly oriented honeycomb in the absence of gravity to guide the construction?" The answer was yes.

Honeycombs are all-purpose housing, storage vessels, and behavioral platforms for the bees. Nectar is a sticky liquid that readily adheres to, or is absorbed into, materials, including most paper. The waxen cells of honeycombs are perfect for storing nectar, which can neither penetrate nor flow along the wax surface. Honeycomb cells are also ideal for storing forager-collected pollen for future consumption. Worker and male bees are reared in honeycomb cells. The queen lays an egg in a cell, the egg hatches into a larva, the larva is fed by specialist nurse worker bees, the mature larva pupates in the cell, and, finally, the young adult bee emerges from the cell. The cell is then cleaned and either another cycle is repeated, the cell is used for storage of nectar or pollen, or it is left in reserve for future use. In addition to the basic uses of the honeycomb, it serves as a communication platform and business conference-room table for recruiting other workers to newly discovered flowers or sugar sources.

North American and Western European cultures have traditionally viewed honey bees as a source of honey for consumption and beeswax for candles and artistic uses. In most of the rest of the world, honey bees are valued for much more than for simple consumption of honey and use of beeswax. Honey bees are valued for their nutritious protein-, vitamin-, and mineral-rich bee brood and pollen, for their propolis for medicinal and sanitation uses, for bee venom for its health and curative value, and even royal jelly—that food of queens—for beauty and possibly health reasons.[1] Most human hunter-gatherer societies highly value bee combs as a nutritional favorite derived from their rich abundance of bee brood, honey, and pollen.[2] Propolis, a plant resin collected by foraging bees, has antimicrobial, antiviral, and antifungal properties. Traditionally, it has found use in treating mouth, gum, and throat disorders; in healing wounds; and as a local surgical anesthetic. In some eye-surgery studies, propolis was found to be three times as powerful as cocaine, an old standard, and 52 times that of procaine for anesthesia.[1] Bee stings and venom have found wide use around the world for treating rheumatoid arthritis and other autoimmune disorders. Young, and not-so-young, women, especially in East Asia, have valued royal jelly for its alleged beauty-enhancing properties in cosmetics and nutritional supplements. Honey has traditionally played important roles in human health, especially in preventing infections in wounds, promoting wound healing, treating severe burns, and as an excellent treatment for slow-healing and ulcerative wounds. Recently, even in the United States, a wound treatment called "MediHoney" has come into use for wound treatment. MediHoney is honey from New Zealand mānuka flowers, a species shown to produce honey highly effective in treating wounds. Maybe we should expand the old saying about pigs that "we use everything except the squeal" to honey bees that "we use everything, except the buzz."

In North America and Europe, the term "honey bee" usually means one particular species of honey bee, the western hive bee, *Apis mellifera*. The western hive bee is but one species in the genus *Apis*, a collection of nine commonly recognized species plus numerous subspecies, or races, within these species. Most of these, including the giant honey bees, the dwarf honey bees, and the eastern hive bees, are found in southern Asia. Often three or more species live in a region where they partition resources according to their body sizes. Giant and dwarf honey bees make nests consisting of one comb that hangs in the open. The hive bees, as the name suggests, live in cavities, often in trees, thus forming hives of bees. *Apis dorsata* is one of the two giant honey bees famous for their impressive defensive attacks against predators. They generally build their single comb as large as 1.5 meters in length and 0.9 meters downward under high branches of tall trees. Frequently, several to as many as 156 colonies will aggregate in a single tree or adjoining trees. When threatened, giant honey bees, unlike hive bees that must exit through a restricting entrance, can simply fall off the comb and instantly mount immense stinging attacks. With an average population of 15,000–40,000 bees per colony, that makes a lot of bees on the attack. If other nearby colonies are disturbed, that attack can be multiplied by a large number. No wonder Roger Morse, one of the great bee scientists of the second half of the twentieth century, commented, "There is no question that *A. dorsata* is the most ferocious stinging insect on earth," a statement few would dispute.[3]

My own first experience with giant honey bees occurred in Borneo with my colleague Chris Starr and our wives. We were in the town of Kota Kinabalu, near Mt. Kinabalu, a peak nearly 4,100 meters high, towering several thousand meters above the next highest point in Borneo. One of our goals was to climb to the top of the peak, along the way surveying the stinging bees, wasps, and ants of the mountain. Before heading out, we discovered a small swarm of giant honey bees in the backyard of our rented room. We were somewhat ill clad for the operation, with only one full bee suit, one insect net with long extension handles, another net, flashlights, and two green army-type mosquito veils. The night was dark, a definite plus, and with two flashlights shining from opposite sides, I took advantage of the dark area in the middle to scrape the swarm off the mid-level branch with my fully extended net. Success. Well, except for the explosion. Most of the bees were in the net, but the odd hundred or so escapees were rocketing down the light beams toward the two, mostly unprotected, colleagues. Lights off. The attack shifted to me. Fortunately, my bee suit armor held and none of us was stung. Imagine if it had been a full colony of 30,000 bees, rather than a mere "docile" swarm of 1,114 workers plus an odd 171 males, lacking a queen.

Giant honey bees' fame is not limited just to bee scientists and the people living among them; it cuts a much wider and more colorful swath. Public complacency about Southeast Asia was shattered September 13, 1981, in a speech by then Secretary of State Alexander Haig, best known for his ill-conceived speech March 30, 1981, just after President Ronald Reagan was shot, in which he claimed, "I am in control here." In his September speech at a Berlin press conference, Haig stated, "the

Soviet Union and its allies have been using lethal chemical weapons in Laos, Kampuchea and Afghanistan." The lethal chemical weapons fell from the sky in the form of a "yellow rain." The alleged chemical weapons were trichothecene fungal toxins in this yellow rain that poured from the sky onto the Hmong peoples of highland areas of Laos. The rain was alleged to be retaliation for their assistance to the U.S. forces during the Vietnam War. The evidence was a single reported analysis from a Minnesota laboratory of minute quantities of three fungal toxins in samples of yellow rain spots. Never mind that the U.S. Army had analyzed more than 50 samples and turned up nothing. It turned out that the Minnesota lab had inadvertently contaminated the samples with the toxins, which they routinely analyzed in the lab. This mistake was not understood until several years later. In the meantime, a flurry of articles and papers riveted the attention of the public and the scientific communities in local newspapers throughout the nation and in the most prestigious journals of *Nature* and *Science*. Matthew Meselson of Harvard University teamed up with Tom Seeley, then at Yale University, to go to Southeast Asia and study yellow rain directly. Their findings: yellow rain was nothing more than giant honey bee poop. Giant honey bees take daily cleansing flights in which thousands of bees simultaneously leave their combs high in the forest and fly a short distance and defecate, raining yellow droplets that form yellow spots on anything below, including people. Tom and Matthew directly observed these flights, collected samples, and analyzed them in their labs. Both the spots they collected and those provided by the army contained no toxins, only pollen, not a surprising discovery, given that honey bees eat pollen.[4] Despite having no valid evidence of chemical warfare agents in the yellow rain, officials continued to generate wild scenarios to explain their toxic arguments. For these explanations to have been plausible, we would have had to credit the USSR with amazing cleverness and talent beyond our imagination. The final nail in the government's case was delivered in a 1987 report in the journal *Foreign Policy* titled "Yellow Rain: The Story Collapses." Nevertheless, no apologies were issued, and, as of 2012, an army manual still listed yellow rain as a potential weapon.

Eastern hive bees living in Japan, often simply called Japanese honey bees, have their own fascinating stories of warfare to tell. These honey bees, *Apis cerana*, are much smaller than giant honey bees and substantially smaller than the familiar western honey bees that live in white hive boxes throughout North America and Europe. Their war was not against the United States, the Soviet Union, or the Hmong people but against giant Japanese hornets, *Vespa mandarinia*. This behemoth of a wasp is the largest stinging insect on Earth, weighing 2 grams to 3.5 grams. These huge hornets, with their orange blocky heads and powerful stings, are fond of preying on other hornets, wasps, and honey bees. Giant hornets are the "meat heads" of the insect world. Their huge heads are mainly composed of muscles to power giant cutting and crushing jaws. These jaws are used to dispatch prey quickly with crushing bites. Prey wasps have only marginal abilities to defend against a giant hornet attack and often will abandon their nest. Western honey bees, *A. mellifera*, are pathetically helpless when giant hornets confront them. As few as 10 hornets make quick work of thousands of defending bees, simply crushing them at a rate of one every couple of seconds.

When giant hornets attack honey bees, the goal is not the adult bees—they are crunchy, full of chemicals, and have little meat relative to shell—but the succulent bee larvae and pupae. After slaughtering the defending adult bees, the hornets invade the colony, feasting on the bee brood and honey with impunity.[5] The smaller native Japanese honey bee is dwarfed in size by giant hornets, but size is not everything. Japanese honey bees have their own magical solution.

After detecting a scouting hornet, instead of attacking the hornet, outside bee flight action stops. The alerted, defending bees mass at the colony entrance, withdraw a bit, and form a tight cluster. By withdrawing into the hive they entice the hornet to come closer. If a hornet does come too close, a phalanx of hundreds of bees instantly attack, grabbing legs, antennae, wings, and anything suitable to spread eagle and immobilize the hornet. Here's where the real trick comes in. They don't try to sting the hornet, likely a futile attempt. Honey bees can thermoregulate and raise their body temperature. That is how they stay cozy in their hives during Canadian or Northern Japanese winters. Japanese honey bees apply this warming ability plus the generation of metabolic carbon dioxide to heat and poison the hornet at the center of the thick ball of bees. The bees raise the temperature to as high as 45°C to 47°C (127°F) and the CO2 level to 3.6% (about the same as human breath). That combination of temperature and CO2 kills the hornet, but doesn't harm the bees, which can withstand up to 50°C. A few degrees difference in temperature tolerance makes all the difference in the world. The dead hornet is now discarded and the bees, having won the battle, get back to work.[6,7]

Honey bees as stinging warriors also live in Africa. In this case, the bees are the familiar hive bee, *Apis mellifera*. Beekeeping is a time-honored activity in Africa, where bees are kept for their honey, wax, and cultural and medicinal use. African honey bees tend to be feisty and take umbrage at disturbers, a characteristic used for protective advantage in some areas of central Africa. Protection is not from people but from marauding elephants. Elephants have enormous appetites and have developed a fondness for human crops. They break fences, sometimes threaten people and livestock, and by eating the crops, leave people hungry. Elephants are called pachyderms, in reference to their thick skin (in Greek, pachy = thick and derm = skin), but they have chinks in that dermal armor. Honey bees share with farmers a dislike of elephants because elephants tend to eat the trees that bees live in. One of the impressive abilities of honey bees is the knowledge of where an attacker, whether human, bear, or elephant, is vulnerable. In the case of elephants, the vulnerable parts are the eyes and inside of the trunk. These are exactly the spots targeted in the stinging attacks by bees. The result is six tons of elephant fleeing the scene and not coming close again.[8]

African farmers learned that elephants are afraid of honey bees (though not mice as portrayed in cartoons) and use this knowledge to their benefit. Beehives are strategically placed around crops to keep crops, people, and elephants safely apart. Elephants are intelligent and quickly learn to avoid getting close to beehives. In addition, elephants have learned to distinguish the voices of different categories of humans. They show little fear on hearing the voices of women or children or of adult men of the local Kamba people. In contrast, elephants show fear and often retreat on hearing voices of Masai men. Masai tribesmen tend to spear intruding elephants, whereas Kamba men do not. The ability of elephants to recognize threatening sounds has even gone high tech. Small airborne drone planes that emit the sounds of buzzing honey bees have been used to herd elephants that have wandered off their game preserves back onto their preserves.[9]

Nearly everybody has heard of killer bees. Are killer bees, also referred to as Africanized bees, some exotic new species of bee? Who actually are these killer bees? Turns out killer bees are just ordinary honey bees that have an "attitude." They do not like potential predators or intruders and exhibit their displeasure through massive stinging attacks. Perhaps somewhat surprisingly, killer bees are not larger, or necessarily blacker, than familiar docile domestic European hive bees; in fact, they tend to be smaller. What they lack in size they make up in behavior.

Domestic bees are unusual among honey bees of the species *Apis mellifera*. Most of the numerous races of *A. mellifera* are defensive, with the exception of the domestic honey bees, which are more docile. They are more docile because of beekeeper breeding in which defensive colonies tended to be killed or have their queens replaced and gentle colonies were used for stock improvement. A century or so of this selective breeding has produced the docile bees kept in white boxes today. As always, some exceptions can, and do, occur with domestic bees, something not lost on beekeepers. Selection among the ancestors of killer bees was just the opposite. In their case, predators, mainly chimpanzees and humans and their ancestors, honed honey bee defenses to the maximum. Those bees that mounted extraordinary stinging defenses against honey hunters tended to survive more frequently than those mounting weaker defenses. This type of selection pressure over a million or more years resulted in the highly defensive honey bees we see today in Africa.

The saga of killer bees began about 60 years ago. The honey bees originally imported into Brazil from Europe performed poorly in the Brazilian tropical and subtropical environment. The bees produced little honey and were plagued with diseases, predators, and poor survival. In response, the government commissioned Dr. Warwick Kerr, known as "Varvique" to his colleagues and friends, to import some bees more suitable to the Brazilian climate. Kerr, a distinguished and talented bee scientist and geneticist, who is still active at age 93, did just that. He brought in and established in hives 48 queen African honey bees from the area around Pretoria, South Africa, and Tanzania, areas similar to parts of Brazil. They performed well in Brazil. While Kerr was away from the apiary one weekend, a visiting scientist opened the hive entrances, allowing 26 reproductive swarms to escape into the countryside. They thrived and reproduced vigorously in the wild. And they were wild, especially wild in their strong defensive attacks toward predators, people, pets, and livestock. Once liberated, the descendants of this original African bee stock began their rapid territorial expansion and arrived in south Texas in 1990. Throughout this trip northward, the wild African-derived bees moved through tropical areas, where they were highly adapted and displaced the poorly adapted honey bees originally brought in by the Spanish and Portuguese from the much colder Europe. The New World had no native honey bees, thus the honey bees present before the arrival of the bees from Africa were all best suited to European climates, not the warmer climates of much of South America and North America. The new African arrivals had to "fight" their way north, in the process stinging and defending themselves from predators and people unaccustomed to vigorously stinging bees. Consequently, the moving front of bees maintained its hot temper and tendency to attack right to Texas.

Warwick Kerr was politically progressive and an outspoken critic of the military dictatorship that came to rule Brazil after the African bees escaped and became known for their dispositions. In part to discredit Kerr, "his" bees were called *abejas assinados* by the authorities and press. In 1965, this catchy name was seized on by *Time* magazine and translated into English as "killer bees." The name has stuck. Never mind that Kerr was doing his job, did it well, and continued to improve the bee's genetics and management, such that Brazil went from the 27th-largest honey-producing nation in 1970 to the 5th largest in 1992 because of the bees from Africa.

Honey and stings. Those are what come to mind about honey bees. We like honey; it is familiar. We don't like stings and think we know more than we'd like to know about them. The mechanical part of the slender stinger poking into the skin is the trivial part of a sting. It is the injected venom that counts. Honey bee venom is the best known of all insect venoms. Its chemical composition has been intensely studied since the 1950s and consists of two major proteins plus an assortment of other minor components. The main component, a small peptide found nowhere in nature except in the venoms of honey bees, is named melittin, a name derived from the scientific name for the honey bee *Apis mellifera.* Melittin is composed of 26 amino acids and comprises around half or so of the total venom. The first biological activity characterized for melittin was its striking ability to destroy red blood cells; thus it was labeled a "hemolysin" for this hemolytic activity. This became an unfortunate label, as melittin's activity became pigeonholed in the minds of subsequent scientific investigators. In reality, melittin does much more than destroy blood cells: It causes pain and, in fact, is the only component in bee venom that causes the immediate sting pain; it greatly enhances the activity of the second-most abundant venom component, phospholipase, and it is a toxin that directly attacks the heart muscle. In hindsight, melittin would be better characterized as an algogen and cardiotoxin for its profound abilities to cause pain and poison the heart.

The second-largest component of honey bee venom, comprising around 20 percent of the total venom, is the protein enzyme phospholipase A_2. Phospholipase destroys phospholipids, essential components of cellular membranes, in the process releasing lysophospholipids that indirectly initiate a variety of other reactions and contribute to producing minor pain. Phospholipase's activity of attacking membrane phospholipids is greatly enhanced and synergized by trace amounts, even well below 1 percent, of melittin. How much activity phospholipase has in the absence of melittin is unclear.

In addition to melittin and phospholipase in honey bee venom are a gang of minor components, none of which comprises more than 4 percent of the total venom. The two most famous of these are apamin and mast-cell-degranulating peptide. Apamin was given its name from the genus name of honey bees, *Apis* (it couldn't be named after the species name, *mellifera*, because that was already taken), and acts as a neurotoxin. Its only problem is that it acts in mammals mainly on the brain; yet, it is blocked by the brain's protective blood-brain barrier. Thus, apamin in the venom is mostly ineffective against vertebrates. The other notorious component, the awkwardly named mast-cell-degranulating peptide (abbreviated MCD-peptide), is known mostly from its profound ability to cause mast cells in the body to degranulate. Degranulating mast cells release a cocktail of highly active components, including histamine, leukotrienes, cytokines, and a head-spinning number of other components. These components cause skin redness, swelling, rash, and a variety of other symptoms. Again, it is unclear how much this venom component contributes to the immediate sting reaction we experience when stung.

In almost any conversation about honey bee stings, the topic of bee sting allergy is mentioned, often with statements such as "My doctor told me I am deathly allergic to bees and might die if stung again." Technically, this is correct if one considers 1 chance in 60,000 to be "might." Granted, one's chances of dying from a cow falling from the sky are less (a cow was actually recorded once falling out of an airplane) but chances of dying from a lightning strike are greater than from a bee sting. Far too much fear is generated relative to the risks of death from a bee sting.

Even more panic and fear permeate discussions of killer bee attacks. In these cases, the perceived threat is death directly from massive envenomation, not from a "possible" or "likely" lethal allergic reaction. Statistics show the contrary. In the United States, since the arrival of killer bees in 1990, only about six to eight toxic deaths from bee attacks have been carefully documented. The rest were deaths from allergic reactions. Leslie Boyer and I showed that a typical person can withstand six stings per pound of body weight and survive the bee attack without medical assistance. Ten stings per pound, in contrast, is lethal. Thus, a 170-pound person could survive 1,000 stings.[10] If we halve that number to be safe, the person would not be at serious toxic risk with fewer than 500 stings. In contrast, death from an allergic reaction can be caused by 1 sting or by 100 stings. When carefully investigated, the majority of deaths blamed on massive bee attacks are allergic deaths, not toxic deaths. A take-away message for medical-treating personnel is to pay attention to allergic problems in massive bee attacks.

An urgent need for an antivenom to combat toxic envenomations became apparent after the arrival of killer bees in the United States. The idea was to produce an antivenom that would neutralize the venom toxins and save lives, just as antivenoms to snake bites and scorpion stings save lives. Attempts to generate protective antibodies in animals, as with snake venoms, failed with honey bees. Antibodies were nicely raised against the major allergenic components of honey bee venom, including phospholipase and hyaluronidase, but these antibodies did not protect mice from lethal doses of venom. Throughout these investigations, nobody asked how honey bee venom actually killed. My hunch was that melittin, the small peptide that does not readily induce antibody generation, was the major cause. Because animal- or even beekeeper-generated antibodies to melittin were not produced in meaningful levels, the melittin in venom challenges would not be neutralized. The bee venom components melittin, phospholipase, and apamin were individually tested for lethality. Apamin had low lethality and was present in only small amounts in honey bee venom, so it was ruled out as a killing factor. Phospholipase alone was the most lethal bee venom component, but it is present in quantities only about one-third those of melittin.

The answer to who's the guilty party in honey bee venom that kills the victim came from a recombination experiment in which pure melittin and pure phospholipase were recombined in the natural 3:1 ratio found in honey bee venom. The combined two components had the same killing power as the melittin alone. In other words, the phospholipase was contributing no more than any inert protein to the overall lethality and was not enhancing the activity of the melittin. The two were operating independently. Autopsies revealed that phospholipase kills by congesting the lungs with fluid and blood, and melittin kills by stopping the heartbeat. The combination of the two stopped the heart and left the lungs clear.[11] The answer to what kills in honey bee venom is melittin, and because it could not be neutralized by current antivenoms that lack antibodies to melittin, the antivenom did not work.

With the arrival of the new bee on the block came questions about how the stings and venom of this new bee differed from those of our ordinary honey bees. Emotionally satisfying expectations were that killer bee stings and venom would hurt more and be more toxic than ordinary familiar bees. In contrast, honey bee defenders claimed with an air of confidence that the two were the same. These statements were simply wishful thinking based on no evidence whatsoever. With the aid of my colleagues Michael Schumacher, an allergist, and Ned Egen, a bioengineer, we decided to investigate these questions. The venoms of killer and domestic bees were similar, differing mainly in the relative ratios of melittin and phospholipase, and had the identical LD_{50} values to mice.[12] Perhaps counterintuitively, domestic honey bee stings hurt more. The reason appeared to be because killer bee venom contained less melittin, the pain-inducing component, than domestic bee venom, and not because killer bees contained less venom. Although they are smaller, killer bees produce about the same quantity of venom as domestic bees. When analyses were extended to other species of honey bees, the results revealed that the venoms of giant honey bees, eastern hive bees, dwarf honey bees, and three strikingly different races of western hive bees all had identical lethalities to mice.[11] The venoms mainly differed in the amount produced by the various bees, with giant honey bees producing eight times as much venom as dwarf honey bees. So, in the end, the venoms of all honey bee species are extraordinarily similar and the experts guessed right that the stings and venoms of killer and domestic bees are the same.

I don't recall my first honey bee sting. I also don't know how many honey bee stings I have received. The number is probably about a thousand, a number seemingly low for someone who has specialized in killer bees for a quarter century. The reasons for the low number are that bee stings are boring, I don't like being stung, and I take precautions. Why boring? Just as eating lots of Halloween candy gets boring after a few days, getting stung by the same species gets boring after a while. The greatest number of stings I ever received at a time occurred from being too casual during a beekeeping operation. I and a helper were moving hives by picking up each hive and carrying it to the new location a few meters away. We had on bee suits and veils but left off the heavy bee gloves as they impeded dexterity. Big mistake. While we were moving a hive, the bottom fell off, just as we got it to carrying height. The whole hive was at risk of falling to the ground and disintegrating. We grabbed the bottom of the exposed hive body to lift it. Unfortunately, about a hundred bees were in a cluster between the hive and my left hand. Squish, sting. Many stings. About 50 in all. We saved the hive. Yes, the stings hurt, but not enough to drop the hive. After five minutes of language best not heard by young children, all was well, save the swelling of the hand for the next day.

My scariest experience with stinging bees occurred in Costa Rica with newly arrived killer bees. I was with my technician who had grown up in a beekeeping family that also operated a beekeeping supply store and beyond doubt had experience, skill, and confidence surpassing that of anybody I knew. We had on bee suits. The day was not optimal. A storm was brewing on the horizon. As we approached within 25 meters, a batch of no-nonsense bees greeted us. A few breached Steve's armor and got inside his veil. In the process of trying to dislodge those bees, he knocked his pith helmet akilter and more bees entered. He panicked and fled. Then, not knowing what to do, I fled right behind him. Lesson learned—don't ever work killer bees with a two-piece veil consisting of hat and veil; always use a one-piece veil, preferably one without a hat that can get bumped.

I am often asked what my worst honey bee sting was. Until recently, my answer was "a sting to the nose or upper lip." A peculiar response to a sting to the nose is that it always seems to cause a series of sneezes. Nobody knows why. Perhaps to sneeze out a bee inside the nose. Stings to the nose or lips really hurt. To make matters worse, stings to the lip invariably swell, irrespective of any "allergy." My most humorous experience (to my colleagues, not to me) occurred in Costa Rica. We came across a nest of a social wasp in the genus *Polybia,* known for its defensiveness. About 10 centimeters away on the same branch as the *Polybia* nest was a small three-wasp nest of *Mischocyttarus.* The little nest was built there to gain benefit from the protection of the nearby "big sister" species. I wanted to identify the small wasp by capturing one of the inhabitants of the tiny satellite nest. The objective was to do this while avoiding arousing the big nest. I attempted to suck one of the docile little wasps into my aspirator. Instead, the little rascal flew off the nest and stung the right side of my upper lip. That night at dinner I was kidded for having a fat right upper lip. The next day it was back to the honey bees, only this time a bee got into my bee veil and stung the left side of my upper lip. At dinner, the chiding was that now I made my swollen lip symmetrical.

Back to my worst honey bee sting. The actual worst sting occurred while I was innocently riding a tandem bicycle with my wife. My mouth was open to get more air. In flew a honey bee and stung my tongue. Extreme pain. Pain much worse than biting one's tongue. This pain truly hurt. It really hurt. Far more than any other honey bee sting. I had to stop, get off the bicycle, and sit on a rock, face in hands. Three minutes of eternity later I could sort of resume riding. Lesson: keep your mouth shut while riding a bicycle.

The difference in sting pain, depending on sting location, is one reason the sting pain scale has only four levels. A superficial honey bee sting to the back of the hand might only rate a 1.5, whereas a sting to the tongue might be a 3. Overall, when combining the values for different locations, an average value of 2 is obtained.

Michael Smith, a Cornell University graduate student, noticed my comments about the differences in the intensity of pain, depending on sting location, and decided to test the subject more thoroughly. In my own experiences, I had simply recorded the pain level and noted the location of the sting. I had never stung myself in various locations or systematically designed stings. I simply recorded what came naturally. Michael, a tall, lean fellow with fluffy red hair and a sense of humor, decided to systematically test pain levels of honey bee stings placed in a random order on 25 different body locations. To increase precision and control variables of bee age and amount of venom delivered, he took bees from the same cohort of guard bees from the hive entrance, placed them on the indicated location, allowed them to sting for 60 seconds, and recorded the pain on a scale of 1 to 10. He received three test stings plus two calibration stings daily over 6 weeks until he accumulated a total of three stings to each location. The chosen locations included some expected locations, such as to the upper arm, forearm, wrist, middle finger, thigh, calf, top of foot, middle toe, lower back, neck, skull, upper lip, and nose. Also included were some nontraditional locations, including buttocks, nipple, scrotum, and penis. As one might well imagine, the latter locations received much public attention. Michael's results ranged from pain ratings of 2.3 to 9.0 and were much as I had expected and experienced. The lowest pain ratings were toes, upper arm, and various other locations on arms and legs. Not surprisingly nose, upper lip, and palm of hand were among the highest pain inducers. The taboo sites were right up there near the top, with the exception of nipple, which was one-third the way down.[13] Michael's study took honey bee sting pain science to a new high.

Honey Bees and Humans: An Evolutionary Symbiosis

A SYMBIOSIS IS A relationship between two different types of organisms that, on the whole, benefits both members. We think of dogs and people, humans and sheep, and some bacteria in our gut as symbiotic. Dogs provide early warning of threats, defend us, help us herd our sheep, and keep us company. We benefit dogs by feeding them, giving them a home, and protecting them. On balance, both dogs and people benefit. The same goes for the sheep-human relationship: sheep provide us wool and meat; we provide protection and pastures for sheep. In our digestive systems are bacteria that synthesize vitamin K. In trade for receiving valuable vitamin K, we provide the bacteria food and a nice home. Symbioses not involving a human are common in nature. A classic example is bees and flowers. The plants benefit from sex, that is, getting sex via bees transporting male pollen to female floral stigmas. The bees benefit by getting food and sometimes other resources. We do not usually think of symbioses as involving stinging insects, other than bees and flowers. For example, we do not think of fire ants as friends or symbionts of us (or anything else, save aphids and other honeydew producers). One exception to our view of symbiosis between stinging insects and another species springs to mind. That exception is the mutually beneficial relationship between bullhorn acacia ants and acacia plants. The plants provide homes in the form of swollen thorns for the ants and food bodies that the ants eat. The ants provide protection from herbivores, such as cows, caterpillars, or leaf beetles and competing plants by stinging herbivores and chewing off plant competitors.

In discussions of symbiosis, I outlined in 2014 one important symbiotic relationship involving our species that has been overlooked.[14]

It is one of our most important symbioses: the symbiosis between humans and honey bees. We think of honey bees as our friends (they provide sweet honey) and our enemies (they sting), but we don't think of us and honey bees as joined at the hip. And, indeed, we are not. We do, however, have a long and colorful history together. The relationship goes back millions of years and involves honey bees and primates, especially chimpanzees and our own ancestors.[2] What both parties have in common is a love of honey. Bees love honey as their food for energy. We love honey for its sweetness and energy. In essence, honey bees are a special example of the arms race between defending stinging insects and attacking predators. Honey bees have outraced many potential nest predators, including most small entomophagous primates that likely would predate bees were it not for their stings and, hence, no longer attempt to exploit the vast honey and protein reserves. Honey bees also appear to have won against many larger primates, including gorillas, bonobos, and baboons. Ratels, those powerful African relatives of wolverines that are also called honey badgers, are the one major nonprimate predator of honey bees in Africa. Predators were the force driving the evolution of honey bee venom and defensive behavior. In Africa, ratels and non-human apes were the most powerful predatory driving forces in this evolution of honey bee defensive behavior. That is, until their role was replaced by humans. Honey bee resources had been and are today exploited by chimpanzees and humans. As Frank Marlowe, who spent decades among the Hadza honey hunters in Tanzania, frames it, "It is not only easy to imagine that humans were collecting honey long before this [cave paintings of 20,000 years ago], but hard to believe otherwise." Almost certainly, early *Homo* lineages preceding modern humans were exploiting the honey and brood of honey bees for millions of years.

The honey bee–hominid (the term for humans and our closest relatives) relationship is special for two reasons. First, of all social Hymenoptera, honey bees store vastly more resources than any other species. These resources include the protein and fat-rich bounty of larvae, pupae, and pollen, plus, uniquely among animals, a huge quantity of sweet and energy-rich honey. Second, chimpanzees and humans are the most intelligent animals in the bee environment. These combined properties set the stage for an amazing evolutionary battle of titans. The result is that bees possess the most powerful insect-stinging defense on Earth, and humans and chimpanzees possess the most sophisticated means of exploiting bee resources of any predators. These superlatives arose because of extreme selection pressure on both parties. To survive, bees needed to evolve the most effective stinging defenses. To exploit the rich, high-quality protein and energy of the hive, hominids needed to learn ways to breach colony defenses and endure many stings. We have no evidence that humans or chimpanzees have evolved any physiological resistance to bee stings—as the mongoose has to cobra venom—but they appear to have developed psychological resistance to the pain of the stings. In essence, they have broken the pain bluff sting signal. Chimpanzees and humans have come to associate little or no damage with dozens, even hundreds, of stings and simply ignore them, or slap at the stinging bees. Jane Goodall, the young English scientist famous for her singlehanded dedication to personal observations of chimpanzees and the discoverer of chimpanzee use of tools for fishing termites, wrote in 1986: "The raiders simply sat eating honey, surrounded by bees at which they merely slapped from time to time, albeit rather frenziedly. Two small infants, in fact, who were clinging to their mothers, whimpered and burrowed their faces into their mother's breasts. Afterward the females spent some time pulling stings from themselves; one mother was also seen to pull them from her infant during a grooming period."[15] Mind over matter occurs when honey is involved. But size and psychological resistance to bee stings do have limits for chimpanzees and humans. Massive honey bee stinging attacks kill people occasionally and frequently drive off chimpanzees, as Goodall noted: "Nine times, after seizing one or two handfuls of honey, the chimpanzees were driven off by swarms of bees, and nine times they ran off without any honey."[15] Defending honey bees sometimes kill the other two specialist honey bee predators, ratels and honeyguide birds, those fascinating birds that lead humans to beehives.

Evidence of the continuing arms race between honey bees and their predators, including humans and our ancestors, is seen in the venom properties and defensive behaviors of honey bees and in the robbing behaviors of humans and our ancestors. The stinging system of honey bees is unusual among stinging insects in the ease with which the sting apparatus pulls out of the honey bee and remains embedded in the flesh of the target. Numerous adaptations make this possible, including a series of sharp backward-facing barbs on the stinger. The barbs render the sting difficult to remove from the skin. Working in tandem with the barbs are preweakened parts of the sting apparatus that allow the apparatus to tear easily and be removed from the bee's abdomen. The resulting torn-out sting apparatus is an independent, self-contained venom system that includes the venom sac and musculature necessary to deliver the venom, plus the nerve ganglion that operates to control the sting and venom movement. The advantage of such a system is that a predator might quickly and easily remove the relatively large bee from its skin before much venom is delivered, but the predator would have difficulty recognizing or removing the tiny speck of a stinger firmly attached in the flesh. By remaining in the flesh, the stinger can deliver all of the venom, not just a small percentage that might be delivered before the bee is brushed off. Another honey bee adaptation is its sting-based alarm pheromone that activates, orients, and stimulates a massive attack by hundreds or thousands of nestmates and other bees from neighboring colonies. Honey bees also behaviorally attack the eyes and nose/mouth area of an attacker, the two most vulnerable and potentially lethal areas on the predator.[10] A final important bee adaptation is the venom. Honey bee venom is among the most lethal of insect venoms, and it is produced in large quantities. This large amount of venom per bee, plus the large number of stinging bees combine to yield a massive killing power of a colony. For example, if half the workers in a colony of 30,000 bees attack and sting, 820 kilograms of predators would receive sufficient venom to yield a 50 percent chance of death. This large colony lethal capacity is sufficient to threaten the lives of 10 or more adult humans and is one reason most humans do not attempt to harvest bee resources but instead leave the task to specialists among their population.

Adaptations brought into play on the chimpanzee and the human lineage side of the relationship with bees include tool making for breaking into nest cavities, learning, use of a mutualistic relationship between honeyguide birds and humans to locate bee colonies, fire, and, recently, domestication and artificial selection and breeding. The tools used by chimpanzees to gain access to hive and honey consist of various clubs, prying and dipping sticks, and leaf sponges. The early human ancestor *Homo erectus* used a more elaborate set of tools in daily life than chimpanzees and that would probably have allowed them to extract honey and combs more efficiently.[16] Present-day humans use a wide variety of traditional tools for robbing honey bee colonies, including various ladders, ropes, hammers and climbing pegs, harnesses, and vessels for collecting the combs. Modern beekeepers have added to the mix metal hive tools, manufactured bee homes (hive boxes), sting-resistant bee suits, and smokers for generating controlled amounts of smoke. Learning and cultural transmission via both spoken and written means are key to human success in managing and exploiting bee resources. Based on learning and cultural transmission of ant fishing methods in chimpanzees, chimpanzees would also be able to learn and transmit methods of exploiting honey bees.[17]

A unique adaptation altering the relationship between humans and honey bees is the mutualism between the greater honeyguide bird, *Indicator indicator*, and humans. The honeyguide, a small brood-parasitic bird that possesses an unusual ability to digest beeswax, guides human honey hunters to a beehive through a series of elaborate behaviors, including unique calls. The bird benefits in this relationship by obtaining comb and hive scraps left after the honey hunter has opened the beehive cavity and taken honey and brood. This mutualism exists only between humans and the bird, perhaps with an earlier relationship between *H. erectus* and the honeyguide.[16] This unique, adaptive mutualistic relationship between bird and man tips the bee-human, predator-prey relationship toward favoring the human side.

A crucial adaptation in the ongoing arms race between predatory humans (for purposes here, ignoring earlier possibilities involving *H. erectus*) and their honey bee prey was the acquisition and management of fire around 1.8 million years ago. Fire is a normal phenomenon in savanna regions of Africa to which bees adapted by filling their honey stomachs with honey in response to smoke and then abandoning the nest and cavity. Concurrently, smoke naturally reduces the defensiveness and stinging behavior of the bees, setting the stage for early humans to exploit the use of fire as an aid in robbing honey and brood from the bees. In this regard, humans turned the tables on the bees by mimicking wildfire, a real and dangerous threat, and thereby tricking them into adopting a weak defense of their hive or the outright abandonment of the hive with little or no defense.[18] Just as Batesian mimicry of a dangerous animal by a harmless animal allows the "cheating" mimic to gain protective benefit, fire use by humans cheats honey bees of their ability to effectively use their stings to repel the attacker. Chimpanzees, though more able to endure stings than humans, do not match the success of human predators, likely because they did not tame fire and cannot endure the resultant massive number of stings. Evolution of increasingly painful and toxic venom cannot overcome the fire-based and disarming attack strategy of humans. The importance of the use of fire by human populations to gain the upper hand in this arms race can hardly be overstated and has resulted in major changes in human evolution.[16]

The most recent addition of humans to their arsenal of weapons for use against honey bees is domestication through artificial selection of bees. From the discussion above, one might conclude that bees are ultimately losing the arms race to humans and might be on a path to extinction. But, instead, by possessing resources highly sought by humans, and by pollinating many human food crops, honey bees have become indispensable mutualists with humans. Humans learned how to manipulate honey bee genetics to reduce their normal defensiveness and to increase their investment in nectar collection and honey production compared to their investment in reproduction. In essence, humans generated a domesticated form of honey bees. This mutualism, like mutualisms in general, benefits both parties overall, even though the parties retain conflicting interests. In trade for a reduction in their tendency to sting and for increased tolerance in having their resources robbed, bees gained reduction in being killed, protection from other predators, and, most important, active dispersal by man from their original habitats of Africa and Europe to all inhabitable parts of the world. The sting set the stage for these benefits for bees by enabling them to defend against large predators and, thereby, to store large quantities of brood, pollen, and honey. These large stores, in turn, first enticed humans to predate bees and then, ultimately to protect, to maintain, and to disperse them throughout the world. In the end, a fine symbiosis was established.

APPENDIX

PAIN SCALE FOR STINGING INSECTS

Note: CA = Central America; NA = North America; SA = South America.

NAME	RANGE	DESCRIPTION	PAIN LEVEL
ANTS			
Indian jumping ant *Harpegnathos saltator*	Asia	Ah, that wonderful wake-up feeling, like coffee but oh so bitter.	1
Delicate trap-jaw ant *Anochetus inermis*	SA	A tiny spark, just enough to rouse you from a dreamy stroll in the woods. A slight jolt as you return to reality.	1
Bothroponera striglosa (A type of African black ant)	Africa	Timid but not painless. A pebble kicked up by a passing car ricochets off your ankle.	1
Asian needle ant *Brachyponera chinensis*	Native to Asia	Nightfall following a day at the beach. You forgot the sunscreen. Your burned nose lets you know.	1
Big-eyed ant *Opthalmopone berthoudi*	Africa	A sharp interruption as you absorb the beauty of Africa. An acacia spine just poked through your sandal.	1
Ectatomma ruidum (A type of black ant)	CA & SA	A brief searing, like tuna on a grill. The bottom of your foot is blanched but not cooked through.	1
Leptogenes kitteli (A type of Asian army ant)	Asia	Simple and plain. A loose carpet tack pierces the ball of your wool-socked foot.	1
Elongate twig ant *Pseudomyrmex gracilis*	NA, CA & SA	Reminiscent of a childhood bully. Intimidating, but his punch only glanced your chin, and you live for another day.	1

NAME	RANGE	DESCRIPTION	PAIN LEVEL
Slender twig ant *Tetraponera* sp.	Asia	A skinny bully's punch. It's too weak to hurt but you suspect a cheap trick might be coming.	1
Red fire ant *Solenopsis invicta*	Native to SA	Sharp, sudden, mildly alarming. Like walking across a shag carpet and reaching for the light switch.	1
Tropical fire ant *Solenopsis geminata*	Native to CA & SA	You should have learned, but the carpet is the same, and when you again reach for the light switch, the shock mocks you.	1
Southern fire ant *Solenopsis xyloni*	NA	It happens on the third day, as you reach for the light switch, and you're wondering when you will ever learn.	1
European fire ant *Myrmica rubra*	Native to Europe	The prickle of stinging nettles against your skin on a hot, humid day.	1
Samsum ant *Euponera sennaarensis*	Africa	Pure, sharp, piercing pain. You pressed your thumb on a tack.	1.5
Suturing army ant *Eciton burchellii*	CA & SA	A cut on your elbow, stitched with a rusty needle.	1.5
Ectatomma tuberculatum (A large golden ant)	CA & SA	A slow stream of hot wax poured over your wrist. You want to twist away but can't.	1.5
Giant ant *Dinoponera gigantea*	SA	A pulsing sting with some flavor. You stepped into a salt bath with an open wound.	1.5

NAME	RANGE	DESCRIPTION	PAIN LEVEL
Giant stink ant *Paltothyreus tarsatus*	Africa	You must have upset the nurses. They stuck you with a big needle and then dribbled in garlic oil.	1.5
Bulldog ant #1 *Myrmecia simillima*	Australia	Intense, ripping, and sharp. The dog's tooth found its mark.	1.5
Red bull ant *Myrmecia gulosa*	Australia	A sneaky, unassuming ache. Like a brightly colored LEGO, charming till it's lodged in the arch of your foot in the dark.	1.5
Bulldog ant #2 *Myrmecia rufinodis*	Australia	Shockingly sharp. A scalpel just lanced your palm.	1.5
Matabele ant *Megaponera analis*	Africa	A child's arrow misses its target and finds its home in your calf.	1.5
Ectatomma quadridens (A type of big black ant)	SA	A burning itch to catch your attention, followed by regret, like ordering spicy chicken wings when you have a mouth sore.	1.5
Bullhorn acacia ant *Pseudomyrmex nigrocinctus*	CA	A rare, piercing, elevated sort of pain. Someone has fired a staple into your cheek.	1.5
Jack jumper ant *Myrmecia pilosula*	Australia	The oven mitt had a hole in it when you pulled the cookies out of the oven.	2
Metallic green ant *Rhytidoponera metallica*	Australia	Deceptively painful. Like biting into a green bell pepper to find it is a habanero chili.	2

NAME	RANGE	DESCRIPTION	PAIN LEVEL
Diacamma sp. (A type of black ant)	Asia	Nothing subtle about this one. Sudden and striking, a shard of glass on a tropical beach pulls you out of the moment as it finds a nerve in your bare foot.	2
Large tropical black ant *Neoponera villosa*	NA, CA & SA	Exquisitely sharp and expertly clean. Broadway's favorite barber selects his next victim.	2
Neoponera crassinoda (A type of big black ant)	SA	This one packs a punch. The dentist should have given that novocaine more time to take effect.	2
Termite-raiding ant *Neoponera commutate*	SA	The debilitating pain of a migraine contained in the tip of your finger. And their queen can sting like her sisters!	2
African giant ant *Streblognathus aethiopicus*	Africa	The white hot burn of a barbecue fork impaling your hand, paired with the grizzle of a serrated knife.	2
Platythyrea lamellose (A purplish ant)	Africa	Relentless prickling throughout your body. Like wearing a wool jumpsuit laced with pine needles and poison ivy.	2
Platythyrea pilosula (A sleek ant)	Africa	A torturous itch and rash with serious lasting power. Should have spent the extra money and paid for a licensed tattoo artist.	2.5
Trap-jaw ant *Odontomachus* spp.	Worldwide Tropics	Instantaneous and excruciating. A rat trap snaps your index fingernail.	2.5

NAME	RANGE	DESCRIPTION	PAIN LEVEL
Florida harvester ant *Pogonomyrmex badius*	NA	Bold and unrelenting. Somebody is using a power drill to excavate your ingrown toenail.	3
Maricopa harvester ant *Pogonomyrmex maricopa*	NA	After eight unrelenting hours of drilling into that ingrown toenail, you find the drill is wedged in the toe.	3
Argentine harvester ant *Ephebomyrmex cunicularius*	SA	A ferocious pang lasting 12 hours or more. Flesh-eating bacteria dissolve your muscles, one by precious one.	3
Bullet ant *Paraponera clavata*	CA & SA	Pure, intense, brilliant pain. Like walking over flaming charcoal with a 3-inch nail embedded in your heel.	4

BEES

NAME	RANGE	DESCRIPTION	PAIN LEVEL
Triepeolus sp. (A type of parasitic bee)	NA	Did I just imagine that? A little scratch that dances with a tickle.	0.5
Anthophorid bee *Emphoropsis pallida*	NA	Almost pleasant, a lover just bit your earlobe a little too hard.	1
Sweat bees *Lasioglossum* spp.	NA	Light and ephemeral, almost fruity. A tiny spark has singed a single hair on your arm.	1
Cactus bee *Diadasia rinconis*	NA	A skewering message: get lost. Surprising, because you did not touch a cactus spine, until you realize it's from a bee.	1
Cuckoo bee *Ericrocis lata*	NA	Touch of fear unrealized. Oh, and I wanted to show how brave I was!	1

NAME	RANGE	DESCRIPTION	PAIN LEVEL
Giant sweat bee *Dieunomia heteropoda*	NA	Size matters but isn't everything. A silver tablespoon drops squarely onto your big toenail, sending you hopping.	1.5
Western honey bee *Apis mellifera*	Native to Africa and Europe	Burning, corrosive, but you can handle it. A flaming match head lands on your arm and is quenched first with lye and then sulfuric acid.	2
Western honey bee *Apis mellifera* (A special case, tongue)	Native to Africa and Europe	It's crawled into your soda can and stings you on the tongue. It's immediate, noisome, visceral, debilitating. For 10 minutes life is not worth living.	3
Bumble bees *Bombus* spp.	NA	Colorful flames. Fireworks land on your arm.	2
California carpenter bee *Xylocopa californica*	NA	Swift, sharp, and decisive. Your fingertip has been slammed by a car door.	2
Giant Bornean carpenter bee *Xylocopa* sp.	Asia	Electrifying, sharp, and piercing. Next time hire an electrician.	2.5

WASPS

NAME	RANGE	DESCRIPTION	PAIN LEVEL
Club-horned wasp *Sapyga pumila*	NA	Disappointing. A paperclip falls on your bare foot.	0.5
Potter wasp *Eumeninae* sp.	NA	Looks deceive. Rich and full-bodied in appearance but flavorless.	0.5
Little wasp *Polybia occidentalis*	CA & SA	Sharp meets spice. A slender cactus spine brushed a Buffalo wing before it poked your arm.	1

NAME	RANGE	DESCRIPTION	PAIN LEVEL
Great black wasp *Sphex pensylvanicus*	NA	Simple and presumptuous. Your younger sibling just nipped at your pinkie finger.	1
Iridescent cockroach hunter *Chlorion cyaneum*	NA	Itchy with a hint of sharpness. A single stinging nettle pricked your hand.	1
Scarab hunter wasp *Triscolia ardens*	NA	Like a sip of tannin, bitterness lingers.	1
Water-walking wasp *Euodynerus crypticus*	NA	Clever, but trivial? A little like magic in that you cannot quite figure out the difference between pain and illusion.	1
Mud dauber *Sceliphron caementarium*	Native to NA	Sharp with a flare of heat. Jalapeño cheese when you were expecting Havarti.	1
Little white velvet ant *Dasymutilla thetis*	NA	Deceptive, like the name. Immediate, rashy, you want to scratch away the kiss of itch. A sand crab pinched your toe while you tanned.	1
Pacific cicada killer *Sphecius convallis*	NA	Clean. Concentrated dish detergent seeps into a freshly cut finger.	1
Western cicada killer *Sphecius grandis*	NA	Pain at first sight. Like poison oak, the more you rub, the worse it gets.	1.5
Eumeninae sp. (A yellow potter wasp)	NA	A surprising touch of nasty. Like a thorn hidden on the back of a rose stem as you clutch the bouquet.	1.5
Mutillidae sp. (A nocturnal velvet ant)	NA	Itch, burn, and more itch. A toothpick dipped in both itch powder and hot sauce is stuck in your thigh.	1.5

NAME	RANGE	DESCRIPTION	PAIN LEVEL
Paper wasp *Polistes versicolor*	CA & SA	Burning, throbbing, and lonely. A single drop of superheated frying oil landed on your arm.	1.5
Thread-waisted paper wasp *Belonogaster* sp.	Africa	Attention getting. Like the time your classmate stabbed you with a pencil point.	1.5
Ferocious polybia wasp *Polybia rejecta*	CA & SA	Like a trick gone wrong. Your posterior is a target for a BB gun. Bull's-eye, over and over.	1.5
Baldfaced hornet *Dolichovespula maculata*	NA	Rich, hearty, slightly crunchy. Similar to getting your hand mashed in a revolving door.	2
Mischocyttarus sp. (A type of paper wasp)	NA, CA & SA	Robust, a full-bodied wake-up call. Imagine a pair of pliers latched onto your upper lip.	2
Colonial thread-waisted *Belonogaster juncea colonialis*	Africa	Tenacious in a stringy way. Tangy. You can't extract yourself from the man-o-war's tentacles.	2
Unstable paper wasp *Polistes instabilis*	CA	Like a dinner guest who stays much too long, the pain drones on. A hot Dutch oven lands on your hand and you can't get it off.	2
Honey wasp *Brachygastra mellifica*	NA & CA	Spicy, blistering. A cotton swab dipped in habanero sauce has been pushed up your nose.	2
Artistic wasp *Parachartergus fraternus*	CA & SA	Pure, then messy, then corrosive. Love and marriage followed by divorce.	2

NAME	RANGE	DESCRIPTION	PAIN LEVEL
Western yellowjacket *Vespula pensylvanica*	NA	Hot and smoky, almost irreverent. Imagine W. C. Fields extinguishing a cigar on your tongue.	2
Glorious velvet ant *Dasymutilla gloriosa*	NA	Instantaneous, like the surprise of being stabbed. Is this what shrapnel feels like?	2
Nocturnal hornet *Provespa* sp.	Asia	Rude, insulting. An ember from your campfire is glued to your forearm.	2.5
Golden paper wasp *Polistes aurifer*	NA & CA	Sharp, piercing, and immediate. You know what cattle feel when they are branded.	2.5
Yellow fire wasp *Agelaia myrmecophila*	CA & SA	An odd, distressing pain. Tiny blowtorches kiss your arms and legs.	2.5
Fierce black polybia wasp *Polybia simillima*	CA	A ritual gone wrong, satanic. The gas lamp in the old church explodes in your face when you light it.	2.5
Giant paper wasp *Megapolistes* sp.	New Guinea	There are gods, and they do throw thunderbolts. Poseidon just rammed his trident into your breast.	3
Red paper wasp *Polistes canadensis*	CA	Caustic and burning, with a distinctly bitter aftertaste. Like spilling a beaker of hydrochloric acid on a paper cut.	3
Red-headed paper wasp *Polistes erythrocephalis*	CA & SA	Immediate, irrationally intense, and unrelenting. This is the closest you will come to seeing the blue of a flame from within the fire.	3

NAME	RANGE	DESCRIPTION	PAIN LEVEL
Dasymutilla klugii (A huge velvet ant)	NA	Explosive and long lasting, you sound insane as you scream. Hot oil from the deep fryer spilling over your entire hand.	3
Tarantula hawk *Pepsis* spp.	NA, CA & SA	Blinding, fierce, shockingly electric. A running hair dryer has just been dropped into your bubble bath.	4
Warrior (or armadillo) wasp *Synoeca septentrionalis*	CA & SA	Torture. You are chained in the flow of an active volcano. Why did I start this list?	4

REFERENCES

CHAPTER 1. STUNG

General interest reference:

Hrdy SB. 2011. *Mothers and Others: The Evolutionary Origins of Mutual Understanding.* Cambridge, MA: Harvard Univ. Press.

1. Van Le Q, LA Isbell et al. 2013. Pulvinar neurons reveal neurobiological evidence of past selection for rapid detection of snakes. *PNAS* 110: 19000–19005.

2. New JJ and TC German. 2015. Spiders at the cocktail party: An ancestral threat that surmounts inattentional blindness. *Evol. Human Behav.* 36: 163–73.

3. LoBue V, DH Rakison, and JS DeLoache. 2010. Threat perception across the life span: Evidence for multiple converging pathways. *Psychol. Sci.* 19: 375–79.

CHAPTER 2. THE STINGER

General interest reference:

Grissell E. 2010. *Bees, Wasps, and Ants.* Portland, OR: Timber Press.

1. Vollrath F and I Douglas-Hamilton. 2002. African bees to control African elephants. *Naturwissenschaften* 89: 508–11.

2. Starr CK. 1990. Holding the fort: Colony defense in some primitively social wasps. In: *Insect Defenses* (DL Evans and JO Schmidt, eds.), pp. 421–63. Albany: State Univ. New York Press.

3. Smith EL. 1970. Evolutionary morphology of the external insect genitalia. 2. Hymenoptera. *Ann. Entomol. Soc. Am.* 63: 1–27.

4. Schmidt PJ, WC Sherbrooke, and JO Schmidt. 1989. The detoxification of ant (*Pogonomyrmex*) venom by a blood factor in horned lizards (*Phrynosoma*). *Copeia* 1989: 603–7.

CHAPTER 3. THE FIRST STINGING INSECTS

General interest reference:

Evans DL and JO Schmidt, eds. 1990. *Insect Defenses.* Albany: State Univ. New York Press.

1. Brower LP, WN Ryerson et al. 1968. Ecological chemistry and the palatability spectrum. *Science* 161: 1349–50.

2. Hölldobler B and EO Wilson. 2009. *The Superorganism.* New York: Norton.

CHAPTER 4. THE PAIN TRUTH

General interest reference:

Schmidt JO. 2008. Venoms and toxins in insects. In: *Encyclopedia of Entomology,* 2nd ed. (JL Capinera, ed.), pp. 4076–89. Heidelberg, Germany: Springer.

1. Roberson DP, S Gudes et al. 2013. Activity-dependent silencing reveals functionally distinct itch-generating sensory neurons. *Nat. Neurosci.* 16: 910–18.

2. Kingdon J. 1977. *East African Mammals,* vol. 3, Part A. London: Academic Press.

CHAPTER 5. STING SCIENCE

General interest reference:

Evans DL and JO Schmidt, eds. 1990. *Insect Defenses.* Albany: State Univ. NY Press.

1. Schmidt JO. 2015. Allergy to venomous insects. In: *The Hive and the Honey Bee* (J Graham, ed.), pp. 906–52. Hamilton, IL: Dadant and Sons.

2. Aili SR, A Touchard et al. 2014. Diversity of peptide toxins from stinging ant venoms. *Toxicon* 92: 166–78.

3. Hamilton WD, R Axelrod, and R Tanese. 1990. Sexual reproduction as an adaption to resist parasites (a review). *PNAS* 87: 3566–73.

4. Schmidt JO. 2014. Evolutionary responses of solitary and social Hymenoptera to predation by primates and overwhelmingly powerful vertebrate predators. *J. Human Evol.* 71: 12–19.

CHAPTER 6. SWEAT BEES AND FIRE ANTS

General interest references for sweat bees:

Michener CD. 1974. *The Social Behavior of the Bees.* Cambridge, MA: Harvard Univ. Press.

Michener CD. 2007. *The Bees of the World,* 2nd ed. Baltimore: Johns Hopkins Univ. Press.

1. Danforth BN, S Sipes et al. 2006. The history of early bee diversification based on five genes plus morphology. *PNAS* 103: 15118–23.

2. Duffield RM, A Fernandes et al. 1981. Macrocyclic lactones and isopentenyl esters in the Dufour's gland secretion of halictine bees (Hymenoptera: Halictidae). *J. Chem. Ecol.* 7: 319–31.

3. Dufour L. 1835. Etude entomologiques VII Hymenopteres. *Ann. Soc. Entomol. France* 4: 594–607.

4. Barrows EM. 1974. Aggregation behavior and responses to sodium chloride in females of a solitary bee, *Augochlora pura* (Hymenoptera; Halictidae). *Fla. Entomol.* 57: 189–93.

5. Schmidt JO. 2014. Evolutionary responses of solitary and social Hymenoptera to predation by primates and overwhelmingly powerful vertebrate predators. *J. Human Evol.* 71: 12–19.

Fire ants:

1. Tschinkel WR. 2006. *The Fire Ants.* Cambridge, MA: Harvard Univ. Press.

2. Wheeler WM. 1910. *Ants: Their Structure, Development and Behavior.* New York: Columbia Univ. Press.

3. Snelling RR. 1963. The United States species of fire ants of the genus *Solenopsis,* subgenus *Solenopsis* Westwood, with synonymy of *Solenopsis aurea* Wheeler (Hymenoptera: Formicidae). *Bureau Entomol. Calif. Dept. Agr. Occasional Pap.,* no. 3: 1–15.

4. Smith JD and EB Smith. 1971. Multiple fire ant stings a complication of alcoholism. *Arch. Dermatol.* 103: 438–41.

5. DeShazo RD, BT Butcher, and WA Banks. 1990. Reactions to the stings of the imported fire ant. *N. Engl. J. Med.* 323: 462–66.

6. Sonnett PE. 1967. Fire ant venom: Synthesis of a reported component of solenamine. *Science* 156: 1759–60.

7. MacConnell JG, MS Blum, and HM Fales. 1970. Alkaloid and fire ant venom: Identification and synthesis. *Science* 168: 840–41.

8. MacConnell JG, MS Blum et al. 1976. Fire ant venoms: Chemotaxonomic correlations with alkaloidal compositions. *Toxicon* 14: 69–78.

CHAPTER 7. YELLOWJACKETS AND WASPS

General interest references:

Edwards R. 1980. *Social Wasps.* West Sussex, UK: Rentokil.

Evans HE and MJ West-Eberhard. 1970. *The Wasps.* Ann Arbor: Univ. Michigan Press.

Schmidt JO. 2009. Wasps. In: *Encyclopedia of Insects,* 2nd ed. (VH Resh and RT Cardé, eds.), pp. 1037–41. San Diego, CA: Academic Press.

1. Wickler W. 1968. *Mimicry in Plants and Animals.* New York: McGraw-Hill.

2. Bequaert J. 1931. A tentative synopsis of the hornets and yellow-jackets (Vespinae; Hymenoptera) of America. *Entomol. Am.* 12: 71–138.

3. Ross KG and JM Carpenter. 1991. Population genetic structure, relatedness, and breeding systems. In: *The Social Biology of Wasps* (KG Ross and RW Matthews, eds.), pp. 451–79. Ithaca, NY: Cornell Univ. Press.

4. Stein KJ, RD Fell, and GI Holtzman.1996. Sperm use dynamics of the baldfaced hornet (Hymenoptera: Vespidae). *Environ. Entomol.* 25: 1365–70.

5. Schmidt JO, HC Reed, and RD Akre. 1984. Venoms of a parasitic and two nonparasitic species of yellowjackets (Hymenoptera: Vespidae). *J. Kans. Entomol. Soc.* 57: 316–22.

6. MacDonald JF. 1980. Biology, recognition, medical importance and control of Indiana social wasps. *Cooperative Ext. Serv., Purdue Univ.* E-91: 24 pp.

7. Akre RD, WB Hill et al. 1975. Foraging distances of *Vespula pensylvanica* workers (Hymenoptera: Vespidae). *J. Kans. Entomol. Soc.* 48: 12–16.

8. Duncan CD. 1939. A contribution to the biology of North American vespine wasps. *Stanford Univ. Publ. Biol. Sci.* 8(1): 1–272.

9. Madden JL. 1981. Factors influencing the abundance of the European wasp (*Paravespula germanica* [F.]). *J. Aust. Entomol. Soc.* 20: 59–65.

10. Akre RD and JF MacDonald. 1986. Biology, economic importance and control of yellow jackets. In: *Economic Impact and Control of Social Insects* (SB Vinson, ed.), pp. 353–412. New York: Praeger.

11. Phillips J. 1974. The vampire wasps of British Columbia. *Bull. Entomol. Soc. Canada* 6: 134.

12. Jandt JM and RL Jeanne. 2005. German yellowjacket (*Vespula germanica*) foragers use odors inside the nest to find carbohydrate food sources. *Ecology* 111: 641–51.

13. Ross KG and RW Matthews. 1982. Two polygynous overwintered *Vespula squamosa* colonies from the southeastern U.S. (Hymenoptera: Vespidae). *Fla. Entomol.* 65: 176–84.

14. Tissot AN and FA Robinson. 1954. Some unusual insect nests. *Fla. Entomol.* 37: 73–92.

15. Spradbery JP. 1973. *Wasps.* Seattle: Univ. Washington Press.

16. MacDonald JF and RW Matthews. 1981. Nesting biology of the eastern yellowjacket, *Vespula maculifrons* (Hymenoptera: Vespidae). *J. Kans. Entomol. Soc.* 54: 433–57.

17. Schmidt JO and LV Boyer Hassen. 1996. When Africanized bees attack: What you and your clients should know. *Vet. Med.* 91: 923–28.

18. Bigelow NK. 1922. Insect food of the black bear (*Ursus americanus*). *Can. Entomol.* 54: 49–50.

19. Fry CH. 1969. The recognition and treatment of venomous and non-venomous insects by small bee-eaters. *Ibis* 111: 23–29.

20. Rau P. 1930. Behavior notes on the yellow jacket, *Vespa germanica* (Hymen.: Vespidae). *Entomol. News* 41: 185–90.

21. Pack Berisford HD. 1931. Wasps in combat. *Irish Nat. J.* 3: 223–24.

22. Denton SB. 1931. *Vespula maculata* and *Apis mellifica*. *Bull. Brooklyn Entomol. Soc.* 26: 44.

23. Scott H. 1930. A mortal combat between a spider and a wasp. *Entomol. Monthly Mag.* 66: 215.

24. Robbins JM. 1938. Wasp versus dragonfly. *Irish Nat. J.* 7: 10–11.

25. O'Rourke FJ. 1945. Method used by wasps of the genus *Vespa* in killing prey. *Irish Nat. J.* 8: 238–41.

26. Evans HE and MJ West-Eberhard. 1970. *The Wasps*. Ann Arbor: Univ. Michigan Press.

27. Davis HG. 1978. Yellowjacket wasps in urban environments. In: *Perspectives in Urban Entomology* (GW Frankie and CS Koehler, eds.), pp. 163–85. New York: Academic Press.

28. Cohen SG and PJ Bianchini. 1995. Hymenoptera, hypersensitivity, and history. *Ann. Allergy* 174: 120.

29. Schmidt JO. 2015. Allergy to venomous insects. In: *The Hive and the Honey Bee* (J Graham, ed.). pp. 907–52. Hamilton, IL: Dadant and Sons.

30. MacDonald JF, RD Akre et al. 1976. Evaluation of yellowjacket abatement in the United States. *Bull. Entomol. Soc. Am.* 22: 397–401.

31. Grant GD, CJ Rogers et al. 1968. Control of ground-nesting yellowjackets with toxic baits—a five-year testing program. *J. Econ. Entomol.* 61: 1653–56.

32. Wagner RE and DA Reierson. 1969. Yellow jacket control by baiting. 1. Influence of toxicants and attractants on bait acceptance. *J. Econ. Entomol.* 62: 1192–97.

33. Parrish MD and RB Roberts. 1983. Insect growth regulators in baits: Methoprene acceptability to foragers and effect on larval eastern yellowjackets (Hymenoptera: Vespidae). *J. Econ. Entomol.* 76: 109–12.

34. Ross DR, RH Shukle et al. 1984. Meat extracts attractive to scavenger *Vespula* in Eastern North America (Hymenoptera: Vespidae). *J. Econ. Entomol.* 77: 637–42.

35. Reid BL and JF MacDonald. 1986. Influence of meat texture and toxicants upon bait collection by the German yellowjacket (Hymenoptera: Vespidae). *J. Econ. Entomol.* 79: 50–53.

36. Spurr EB. 1995. Protein bait preferences of wasps (*Vespula vulgaris* and *V. germanica*) at Mt Thomas, Canterbury, New Zealand. *N. Z. J. Zool.* 22: 282–89.

37. McGovern TP, HG Davis et al. 1970. Esters highly attractive to *Vespula* spp. *J. Econ. Entomol.* 63: 1534–36.

38. Wildman T. 1770. A treatise on the management of bees. Book 3: *Of Wasps and Hornets and the Means of Destroying Them*, 2nd ed. London: Kingsmeade.

39. Ormerod RL. 1868. *British Social Wasps*. London: Longmans, Green Reader, and Dyer.

40. Rabb RL and FR Lawson. 1957. Some factors influencing the predation of *Polistes* wasps on the tobacco hornworm. *J. Econ. Entomol.* 50: 778–84.

CHAPTER 8. HARVESTER ANTS

General interest references:

Cole AC. 1974. Pogonomyrmex *Harvester Ants*. Knoxville: Univ. Tennessee Press.

Taber SW. 1998. *The World of the Harvester Ants*. College Station: Texas A&M Univ. Press.

1. Creighton WS. 1950. Ants of North America. *Bull. Mus. Comp. Zool. (Harvard)* 104: 1–585.

2. Wheeler WM. 1910. *Ants: Their Structure, Development and Behavior*. New York: Columbia Univ. Press.

3. Lockwood JA. 2009. *Six-Legged Soldiers*. New York: Oxford Univ. Press.

4. Groark KP. 2001. Taxonomic identity of "hallucinogenic" harvester ant (*Pogonomyrmex californicus*) confirmed. *J. Ethnobiol.* 21: 133–44.

5. Blum MS, JR Walker et al. 1958. Chemical, insecticidal, and antibiotic properties of fire ant venom. *Science* 128: 306–7.

6. Herrmann M and S Helms Cahan. 2014. Inter-genomic sexual conflict drives antagonistic coevolution in harvester ants. *Proc. R. Soc. Lond. B Biol. Sci.* 281: 20141771.

7. Johnson RA. 2002. Semi-claustral colony founding in the seed-harvesting ant *Pogonomyrmex californicus*: A comparative analysis of colony founding strategies. *Oecologia* 132: 60–67.

8. Cole BJ. 2009. The ecological setting of social evolution: The demography of ant populations. In: *Organization of Insect Societies* (J Gadau and J Fewell, eds.), pp. 75–104. Cambridge, MA: Harvard Univ. Press.

9. Keeler KH. 1993. Fifteen years of colony dynamics in *Pogonomyrmex occidentalis*, the Western harvester ant in Western Nebraska. *Southwest. Nat.* 38: 286–89.

10. Michener CD. 1942. The history and behavior of a colony of harvester ants. *Sci. Monthly* 55: 248–58.

11. Lavigne RJ. 1969. Bionomics and nest structure of *Pogonomyrmex occidentalis* (Hymenoptera: Formicidae). *Ann. Entomol. Soc. Am.* 62:1166–75.

12. MacKay WP. 1981. A comparison of the nest phenologies of three species of *Pogonomyrmex* harvester ants (Hymenoptera: Formicidae). *Psyche* 88: 25–74.

13. McCook HC. 1907. *Nature's Craftsmen*. New York: Harper & Brothers.

14. Zimmer K and RR Parmenter. 1998. Harvester ants and fire in a desert grassland: Ecological responses of *Pogonomyrmex rugosus* (Hymenoptera: Formicidae) to experimental wildfires in Central New Mexico. *Environ. Entomol.* 27: 282–87.

15. McCook HC. 1879. *The Natural History of the Agricultural Ant of Texas*. Philadelphia: Lippincott's Press.

16. Rogers LE. 1974. Foraging activity of the Western Harvester ant in the shortgrass plains ecosystem. *Environ. Entomol.* 3: 420–24.

17. Knowlton GF. 1938. Horned toads in ant control. *J. Econ. Entomol.* 31: 128.

18. Headlee TJ and GA Dean. 1908. The mound-building prairie ant. *Bull. Kans. State Agr. Exp. Station* 154: 165–80.

19. Clarke WH and PL Comanor. 1975. Removal of annual plants from the desert ecosystem by western harvester ants, *Pogonomyrmex occidentalis*. *Environ. Entomol.* 4: 52–56.

20. Porter SD and CD Jorgensen. 1981. Foragers of the harvester ant, *Pogonomyrmex owyheei*: A disposable caste? *Behav. Ecol. Sociobiol.* 9: 247–56.

21. MacKay WP. 1982. The effect of predation of western widow spiders (Araneae: Theridiidae) on harvester ants (Hymenoptera: Formicidae). *Oecologia* 53: 406–11.

22. Evans HE. 1962. A review of nesting behavior of digger wasps of the genus *Aphilanthops,* with special attention to the mechanics of prey carriage. *Behaviour* 19: 239–60.

23. Knowlton GF, RS Roberts, and SL Wood. 1946. Birds feeding on ants in Utah. *J. Econ. Entomol.* 49: 547–48.

24. Giezentanner KI and WH Clark. 1974. The use of western harvester ant mounds as strutting locations by sage grouse. *Condor* 76: 218–19.

25. Spangler, Hayward G., personal communication.

26. Pianka ER and WS Parker. 1975. Ecology of horned lizards: A review with special reference to *Phrynosoma platyrhinos. Copeia* 1975: 141–62.

27. Schmidt PJ, WC Sherbrooke, and JO Schmidt. 1989. The detoxification of ant (*Pogonomyrmex*) venom by a blood factor in horned lizards (*Phrynosoma*). *Copeia* 1989: 603–7.

28. Schmidt JO and GC Snelling. 2009. *Pogonomyrmex anzensis* Cole: Does an unusual harvester ant species have an unusual venom? *J. Hymenoptera Res.* 18: 322–25.

29. Wray DL. 1938. Notes on the southern harvester ant (*Pogonomyrmex badius* Latr.) in North Carolina. *Ann. Entomol. Soc. Am.* 31: 196–201.

30. Wheeler GC and J Wheeler. 1973. *Ants of Deep Canyon.* Riverside: Univ. California Press.

31. Wray J. 1670. Concerning some uncommon observations and experiments made with an acid juyce to be found in ants. *Philos. Trans. R. Soc. Lond.* 5: 2063–69.

32. Schmidt JO and MS Blum. 1978. A harvester ant venom: Chemistry and pharmacology. *Science* 200: 1064–66.

33. Schmidt JO and MS Blum. 1978. The biochemical constituents of the venom of the harvester ant, *Pogonomyrmex badius. Comp. Biochem. Physiol.* 61C: 239–47.

34. Schmidt JO and MS Blum. 1978. Pharmacological and toxicological properties of harvester ant, *Pogonomyrmex badius,* venom. *Toxicon* 16: 645–51.

35. Piek T, JO Schmidt et al. 1989. Kinins in ant venoms—a comparison with venoms of related Hymenoptera. *Comp. Biochem. Physiol.* 92C: 117–24.

36. Schmidt JO. 2008. Venoms and toxins in insects. In: *Encyclopedia of Entomology*, 2nd ed. (JL Capinera, ed.), pp. 4076–89. Heidelberg, Ger.: Springer.

CHAPTER 9. TARANTULA HAWKS AND SOLITARY WASPS

General interest references:

Evans HE. 1973. *Wasp Farm*. New York: Doubleday.

O'Neill KM. 2001. *Solitary Wasps: Behavior and Natural History*. Ithaca, NY: Cornell Univ. Press.

References for tarantula hawks:

1. Wilson EO. 2012. *The Social Conquest of Earth*. New York: Norton.

2. Swink WG, SM Paiero, and CA Nalepa. 2013. Burprestidae collected as prey by the solitary, ground-nesting philanthine wasp *Cerceris fumipennis* (Hymenoptera: Crabronidae) in North Carolina. *Ann. Entomol. Soc. Am.* 106: 111–16.

3. Sweeney BW and RL Vannote. 1982. Population synchrony in mayflies: A predator satiation hypothesis. *Evolution* 36: 810–21.

4. Hook, Allen W., personal communication.

5. Evans HE. 1968. Studies on Neotropical Pompilidae (Hymenoptera) IV: Examples of dual sex-limited mimicry in *Chirodamus. Psyche* 75: 1–22.

6. Schmidt JO. 2004. Venom and the good life in tarantula hawks (Hymenoptera: Pompilidae): How to eat, not be eaten, and live long. *J. Kans. Entomol. Soc.* 77: 402–13.

7. Pitts JP, MS Wasbauer, and CD von Dohlen. 2006. Preliminary morphological analysis of relationships between the spider wasp subfamilies (Hymenoptera: Pompilidae): Revisiting an old problem. *Zoologica Scripta* 35: 63–84.

8. Williams FX. 1956. Life history studies of *Pepsis* and *Hemipepsis* wasps in California (Hymenoptera, Pompilidae). *Ann. Entomol. Soc. Am.* 49: 447–66.

9. Petrunkevitch A. 1926. Tarantula versus tarantula-hawk: A study of instinct. *J. Exp. Zool.* 45: 367–97.

10. Cazier MA and MA Mortenson. 1964. Bionomical observations on tarantula-hawks and their prey (Hymenoptera: Pompilidae: *Pepsis*). *Ann. Entomol. Soc. Am.* 57: 533–41.

11. Odell GV, CL Ownby et al. 1999. Role of venom citrate. *Toxicon* 37: 407–9.

12. Piek T, JO Schmidt et al. 1989. Kinins in ant venoms—a comparison with venoms of related Hymenoptera. *Comp. Biochem. Physiol.* 92C: 117–24.

13. Leluk J, JO Schmidt, and D Jones. 1989. Comparative studies on the protein composition of hymenopteran venom reservoirs. *Toxicon* 27: 105–14.

References for cicada killers:

1. Rau P and N Rau. 1918. *Wasp Studies Afield.* Princeton, NJ: Princeton Univ. Press.

2. Dambach CA and E Good. 1943. Life history and habits of the cicada killer in Ohio. *Ohio J. Sci.* 43: 32–41.

3. Smith RL and WM Langley. 1978. Cicada stress sound: An assay of its effectiveness as a predator defense mechanism. *Southwest. Nat.* 23: 187–96.

4. Hastings J. 1986. Provisioning by female western cicada killer wasps *Sphecius grandis* (Hymenoptera: Sphecidae): Influence of body size and emergence time on individual provisioning success. *J. Kans. Entomol. Soc.* 59: 262–68.

5. Coelho JR 2011. Effects of prey size and load carriage on the evolution of foraging strategies in wasps. In: *Predation in the Hymenoptera: An Evolutionary Perspective* (C Polidori, ed.), pp. 23–36. Kerala, India: Transworld Research Network.

6. Hastings JM, CW Holliday et al. 2010. Size-specific provisioning by cicada killers, *Sphecius speciosus* (Hymenoptera: Crabronidae) in North Florida. *Fla. Entomol.* 93: 412–21.

7. Alcock J. 1975. The behaviour of western cicada killer males, *Sphecius grandis* (Sphecidae, Hymenoptera). *J. Nat. Hist.* 9: 561–66; and Holliday, Charles H., personal communication.

8. Hastings J. 1989. Protandry in western cicada killer wasps (*Sphecius grandis,* Hymenoptera: Sphecidae): An empirical study of emergence time and mating opportunity. *Behav. Ecol. Sociobiol.* 25: 255–60.

9. Holliday C, J Coelho, and J Hastings. 2010. Conspecific kleptoparasitism in Pacific cicada killers, *Sphecius convallis.* Ent. Soc. Am. Meeting, San Diego, CA [Poster D 0708].

References for mud daubers:

1. Bachleda FL. 2002. *Dangerous Wildlife in California and Nevada: A Guide to Safe Encounters at Home and in the Wild.* Birmingham, AL: Menasha Ridge Press.

2. O'Connor R and W Rosenbrook. 1963. The venom of the mud-dauber wasps. I. *Sceliphron caementarium*: Preliminary separations and free amino acid content. *Can. J. Biochem. Phys.* 41: 1943–48.

3. Frazier C. 1964. Allergic reactions to insect stings: A review of 180 cases. *South. Med. J.* 47: 1028–34.

4. Collinson P. 1745. An account of some very curious wasp nests made of clay in Pensilvania by John Bartram. *Philos. Trans. R. Soc. Lond.* 43: 363–65.

5. Shafer GD. 1949. *The Ways of a Mud Dauber.* Palo Alto, CA: Stanford Univ. Press.

6. Fink T, V Ramalingam et al. 2007. Buzz digging and buzz plastering in the black-and-yellow mud dauber wasp, *Sceliphron caementarium* (Drury). *J. Acoust. Soc. Am.* 122(5, Pt 2): 2947–48.

7. Jackson JT and PG Burchfield. 1975. Nest-site selection of barn swallows in east-central Mississippi. *Am. Midland Nat.* 94: 503–9.

8. Smith KG. 1986. Downy woodpecker feeding on mud-dauber wasp nests. *Southwest. Nat.* 31: 134.

9. Hefetz A and SWT Batra. 1979. Geranyl acetate and 2-decen-1-ol in the cephalic secretion of the solitary wasp *Sceliphron caementarium* (Sphecidae: Hymenoptera). *Experientia* 35: 1138–39.

10. Bohart GE and WP Nye. 1960. Insect pollinators of carrots in Utah. *Utah Agr. Exp. Sta. Bull.* 419: 1–16.

11. Menhinick EF and DA Crossley. 1969. Radiation sensitivity of twelve species of arthropods. *Ann. Entomol. Soc. Am.* 62: 711–17.

12. Muma MH and WF Jeffers. 1945. Studies of the spider prey of several mud-dauber wasps. *Ann. Entomol. Soc. Am.* 38: 245–55.

13. Uma DB and MR Weiss. 2010. Chemical mediation of prey recognition by spider-hunting wasps. *Ethology* 116: 85–95.

14. Uma D, C Durkee et al. 2013. Double deception: Ant-mimicking spiders elude both visually- and chemically-oriented predators. *PLOS One* 8(11): e79660.

15. Konno K, MS Palma et al. 2002. Identification of bradykinins in solitary wasp venoms. *Toxicon* 40: 309–12.

16. Sherman RG. 1978. Insensitivity of the spider heart to solitary wasp venom. *Comp. Biochem. Phys.* 61A: 611–15.

References for iridescent cockroach hunters:

1. Hook AW. 2004. Nesting behavior of *Chlorion cyaneum* (Hymenoptera: Sphecidae), a predator of cockroaches (Blattaria: Polyphagidae). *J. Kans. Entomol. Soc.* 77: 558–64.

2. Peckham DJ and FE Kurczewski. 1978. Nesting behavior of *Chlorion aerarium. Ann. Entomol. Soc. Am.* 71: 758–61.

3. Chapman RN, CE Mickel et al. 1926. Studies in the ecology of sand dune insects. *Ecology* 7: 416–26.

Reference for water-walking wasps:

1. Isely D. 1913. Biology of some Kansas Eumenidae. *Kans. Univ. Sci. Bull.* 7: 231–309.

References for velvet ants:

1. Brothers DJ, G Tschuch, and F Burger. 2000. Associations of mutillid wasps (Hymenoptera, Mutillidae) with eusocial insects. *Insectes Soc.* 47: 201–11.

2. Mickel CE. 1928. Biological and taxonomic investigations on the mutillid wasps. *Bull. U.S. Nat. Mus.* 143: 1–351.

3. Brothers DJ. 1972. Biology and immature stages of *Pseudomethoca f. frigida,* with notes on other species (Hymenoptera: Mutillidae). *Univ. Kans. Sci. Bull.* 50: 1–38.

4. Brothers DJ. 1984. Gregarious parasitoidism in Australian Mutillidae (Hymenoptera). *Aust. Entomol. Mag.* 11: 8–10.

5. Tormos J, JD Asis et al. 2009. The mating behaviour of the velvet ant, *Nemka viduata* (Hymenoptera: Mutillidae). *J. Insect Behav.* 23: 117–27.

6. Brothers DJ. 1989. Alternative life-history styles of mutillid wasps (Insecta, Hymenoptera). In *Alternative Life-History Styles of Animals* (MN Bruton, ed.), pp. 279–91. Dordrecht, Netherlands: Kluwer.

7. Schmidt JO and MS Blum. 1977. Adaptations and responses of *Dasymutilla occidentalis* (Hymenoptera: Mutillidae) to predators. *Entomol. Exp. Appl.* 21: 99–111.

8. Fales HM, TM Jaouni et al. 1980. Mandibular gland allomones of *Dasymutilla occidentalis* and other mutillid wasps. *J. Chem. Ecol.* 6: 895–903.

9. Hale Carpenter GD. 1921. Experiments on the relative edibility of insects, with special reference to their coloration. *Trans. Entomol. Soc. Lond.* 1921: 1–105.

10. Rice ME. 2014. Edward O. Wilson: I was trying to find every kind of ant. *Am. Entomol.* 60: 135–41.

11. Vitt LJ and WE Cooper. 1988. Feeding responses of skinks (*Eumeces laticeps*) to velvet ants (*Dasymutilla occidentalis*). *J. Herpet.* 22: 485–88.

12. Schmidt JO. 2008. Venoms and toxins in insects. In *Encyclopedia of Entomology*, 2nd ed. (JL Capinera, ed.), pp. 4076–89. Heidelberg, Germany: Springer.

13. Schmidt JO, MS Blum, and WL Overal. 1986. Comparative enzymology of venoms from stinging Hymenoptera. *Toxicon* 24: 907–21.

CHAPTER 10. BULLET ANTS

General interest reference:

Young AM and HR Hermann. 1980. Notes on foraging of the giant tropical ant *Paraponera clavata* (Hymenoptera: Formicidae: Ponerinae). *J. Kans. Entomol. Soc.* 53: 35–55.

1. Spruce R. 1908. *Notes of a Botanist on the Amazon and Andes*, Vol. 1, pp. 363–64. London: Macmillan.

2. Lange A. 1914. *The Lower Amazon*. New York: G. P. Putnam's Sons.

3. Rice H. 1914. Further explorations in the north-west Amazon basin. *Geograph. J.* 44: 137–68.

4. Allard HA. 1951. *Dinoponera gigantea* (Perty), a vicious stinging ant. *J. Wash. Acad. Sci.* 41: 88–90.

5. Rice ME. 2015. Terry L. Erwin: She had a black eye and in her arm she held a skunk. *Am. Entomol.* 61: 9–15.

6. Schmidt C. 2013. Molecular phylogenetics of ponerine ants (Hymenoptera: Formicidae: Ponerinae). *Zootaxa* 3647(2): 201–50.

7. Bennett B and MD Breed. 1985. On the association between *Pentaclethra macroloba* (Mimosaceae) and *Paraponera clavata* (Hymenoptera: Formicidae) colonies. *Biotropica* 17: 253–55.

8. Hölldobler B and EO Wilson. 1990. Host tree selection by the Neotropical ant *Paraponera clavata* (Hymenoptera: Formicidae). *Biotropica* 22: 213–14.

9. Belk MC, HL Black, and CD Jorgensen. 1989. Nest tree selectivity by the tropical ant, *Paraponera clavata*. *Biotropica* 21: 173–77.

10. Dyer LA. 2002. A quantification of predation rates, indirect positive effects on plants, and foraging variation of the giant tropical ant, *Paraponera clavata*. *J. Insect Sci.* 2(18): 1–7.

11. Fritz G, A Stanley Rand, and CW dePamphilis. 1981. The aposematically colored frog, *Dendrobates pumilio*, is distasteful to the large, predatory ant *Paraponera clavata*. *Biotropica* 13: 158–59.

12. Harrison JF, JH Fewell et al. 1989. Effects of experience on use of orientation cues in the giant tropical ant. *Anim. Behav.* 37: 869–71.

13. Nelson CR, CD Jorgensen et al. 1991. Maintenance of foraging trails by the giant tropical ant *Paraponera clavata* (Insecta: Formicidae: Ponerinae). *Insect. Sociaux* 38: 221–28.

14. Fewell JH, JF Harrison et al. 1992. Distance effects on resource profitability and recruitment in the giant tropical ant, *Paraponera clavata*. *Oecologia* 92: 542–47.

15. Fewell JH, JF Harrison et al. 1996. Foraging energetics of the ant, *Paraponera clavata*. *Oecologia* 105: 419–27.

16. Jorgensen CD, HL Black, and HR Hermann. 1984. Territorial disputes between colonies of the giant tropical ant *Paraponera clavata* (Hymenoptera: Formicidae: Ponerinae). *J. Ga. Entomol. Soc.* 19: 156–58.

17. Thurber DK, MC Belk et al. 1993. Dispersion and mortality of colonies of the tropical ant *Paraponera clavata*. *Biotropica* 25: 215–21.

18. Barden A. 1943. Food of the basilisk lizard in Panama. *Copeia* 1943: 118–21.

19. Cott HB. 1936. Effectiveness of protective adaptations in the hive bee, illustrated by experiments on the feeding reactions, habit formation, and memory of the common toad (*Bufo bufo bufo*). *J. Zool. Lond.* 1936: 111–33.

20. Janzen DH and CR Carroll. 1983. *Paraponera clavata* (bala, giant tropical ant). In: *Costa Rican Natural History* (DH Janzen, ed.), pp. 752–53. Chicago: Univ. Chicago Press.

21. Brown BV and DH Feener. Behavior and host location cues of *Apocephalus paraponerae* (Diptera: Phoridae), a parasitoid of the giant tropical ant, *Paraponera clavata* (Hymenoptera: Formicidae). *Biotropica* 23: 182–87.

22. Feener DH, LF Jacobs, and JO Schmidt. 1996. Specialized parasitoid attracted to a pheromone of ants. *Anim. Behav.* 51: 61–66.

23. Weber NA. 1937. The sting of an ant. *Am. J. Trop. Med.* 1937: 165–69.

24. Balée W. 2000. Antiquity of traditional ethnobiological knowledge in Amazonia: The Tupí-Guaraní family and time. *Ethnohistory* 47: 399–422.

25. Schmidt JO. 2008. Venoms and toxins in insects. In *Encyclopedia of Entomology*, 2nd ed. (JL Capinera, ed.), pp. 4076–89. Heidelberg, Germany: Springer.

26. Schmidt JO, MS Blum, and WL Overal. 1984. Hemolytic activities of stinging insect venoms. *Arch. Insect Biochem. Physiol.* 1: 155–60.

27. Piek T, A Duval et al. 1991. Poneratoxin, a novel peptide neurotoxin from the venom of the ant, *Paraponera clavata. Comp. Biochem. Physiol.* 99C: 487–95.

CHAPTER 11. HONEY BEES AND HUMANS: AN EVOLUTIONARY SYMBIOSIS

General interest references:

Crane E. 1990. *Bees and Beekeeping*. Ithaca, NY: Cornell Univ. Press.

Graham J, ed. 2015. *The Hive and the Honey Bee*. Hamilton, IL: Dadant & Sons.

Hepburn HR and SE Radloff. 2011. *Honeybees of Asia*. Heidelberg, Germany: Springer.

Wilson-Rich N, K Allin et al. 2014. *The Bee: A Natural History*. Princeton, NJ: Princeton Univ. Press.

1. Schmidt JO and SL Buchmann 1992. Other products of the hive. In: *The Hive and the Honey Bee* (J Graham, ed.), pp. 927–88. Hamilton, IL: Dadant & Sons.

2. Marlowe FW, JC Berbesque et al. 2014. Honey, Hadza, hunter-gatherers, and human evolution. *J. Human Evol.* 71: 119–28.

3. Morse RA and FM Laigo. 1969. *Apis dorsata* in the Philippines. *Monogr. Philippines Assoc. Entomol.*, no. 1: 1–97.

4. Seeley TD, JW Nowicke et al. 1985. Yellow rain. *Sci. Am.* 253(3): 128–37.

5. Matsuura M and SK Sakagami. 1973. A bionomic sketch of the giant hornet, *Vespa mandarinia*, a serious pest for Japanese apiculture. *J. Fac. Sci. Hokkaido Univ. Ser. VI, Zool.* 19: 125–60.

6. Ono M, T Igarashi et al. 1995. Unusual thermal defence by a honeybee against mass attack by hornets. *Nature* 377: 334–36.

7. Sugahara M and F Sakamoto. 2009. Heat and carbon dioxide generated by honeybees jointly act to kill hornets. *Naturwissenschaften* 96: 1133–36.

8. Vollrath F and I Douglas-Hamilton. 2002. African bees to control African elephants. *Naturwissenschaften* 89: 508–11.

9. McComb K, G Shannon et al. 2014. Elephants can determine ethnicity, gender, and age from acoustic cues in human voices. *PNAS* 111: 5433–38.

10. Schmidt JO and LV Boyer Hassen. 1996. When Africanized bees attack: What you and your clients should know. *Vet. Med.* 91: 923–28.

11. Schmidt JO. 1995. Toxinology of the honeybee genus *Apis*. *Toxicon* 33: 917–27.

12. Schumacher MJ, JO Schmidt, and NB Egen. 1989. Lethality of "killer" bee stings. *Nature* 337: 413.

13. Smith ML. 2014. Honey bee sting pain index by body location. *Peer J.* 2:e338; doi:10.7717/peerj.338.

14. Schmidt JO. 2014. Evolutionary responses of solitary and social Hymenoptera to predation by primates and overwhelmingly powerful vertebrate predators. *J. Human Evol.* 71: 12–19.

15. Goodall J. 1986. *The Chimpanzees of Gombe: Patterns of Behavior*. Cambridge, MA: Harvard Univ. Press.

16. Wrangham RW. 2011. Honey and fire in human evolution. In: *Casting the Net Wide: Papers in Honor of Glynn Isaac and His Approach to Human Origins Research* (J Sept and D Pilbeam, eds.), pp. 149–67. Oxford: Oxbow Books.

17. Sanz CM and DB Morgan. 2009. Flexible and persistent tool-using strategies in honey-gathering by wild chimpanzees. *Int. J. Primatol.* 30: 411–27.

18. Buchmann SL. 2005. *Letters from the Hive*. New York: Random House.

INDEX